T0331059

Chemical Ecology

This textbook provides a comprehensive overview of the principles, methods and applications of chemical ecology, covering such topics as chemical signalling, predator–prey interactions, host plant selection and chemical defence. The book takes the reader through the historical development of the discipline to current state-of-the-art research, delving into recent findings on the role of chemical ecology in conservation and management and exploring how the field may contribute to future innovations in ecological science. A chapter is dedicated to the techniques that have been used in chemical ecology and some success stories.

Chemical Ecology: Insect-Plant Interactions is an important resource for advanced undergraduates and postgraduate researchers as well as practitioners in this interdisciplinary field. The book's layout aligns with the curriculum of chemical-ecology–related disciplines, progressing from basic fundamental principles to a more advanced level. Those studying and researching in ecology, entomology, plant biology and biochemistry will find it invaluable as well as those practising in areas such as agriculture, forestry and pest management.

Chemical Ecology

Insect-Plant Interactions

Jamin Ali and Ri Zhao Chen

CRC Press
Taylor & Francis Group
Boca Raton London New York

CRC Press is an imprint of the
Taylor & Francis Group, an **informa** business

First edition published 2025
by CRC Press
2385 NW Executive Center Drive, Suite 320, Boca Raton FL 33431

and by CRC Press
4 Park Square, Milton Park, Abingdon, Oxon, OX14 4RN

CRC Press is an imprint of Taylor & Francis Group, LLC

© 2025 Jamin Ali and Ri Zhao Chen

Library of Congress Cataloging-in-Publication Data
Names: Ali, Jamin, author.
Title: Chemical ecology : insect-plant interactions / Jamin Ali, Foreign
Research Expert, College of Plant Protection, Jilin Agricultural
University, China, and Rizhao Chen, Professor, College of Plant
Protection, Jilin Agricultural University, China.
Description: First edition. | Boca Raton, FL : CRC Press, 2025. |
Includes bibliographical references and index. | Identifiers: LCCN 2024024971 (print) |
LCCN 2024024972 (ebook) | ISBN 9781032767062 (pbk) | ISBN 9781032767338 (hbk) |
ISBN 9781003479857 (ebk)
Subjects: LCSH: Insect-plant relationships. | Chemical ecology. |
Insects--Ecophysiology. | Animal chemical ecology. | Plant chemical ecology.
Classification: LCC QL495 .A45 2025 (print) | LCC QL495 (ebook) |
DDC 591.4--dc23/eng/20240821
LC record available at https://lccn.loc.gov/2024024971
LC ebook record available at https://lccn.loc.gov/2024024972

ISBN: 9781032767338 (hbk)
ISBN: 9781032767062 (pbk)
ISBN: 9781003479857 (ebk)

DOI: 10.1201/9781003479857

Typeset in Minion
by Deanta Global Publishing Services, Chennai, India

Contents

Preface

The realm of chemical ecology has emerged as an enthralling frontier in ecological research, unveiling the intricate dialogues and interactions among organisms through chemical signals. In this book, *Chemical Ecology: Insect-Plant Interactions*, we embark on a comprehensive exploration of this captivating realm, delving into the profound roles that chemical cues play in shaping the relationships and dynamics of the natural world. This handbook is a culmination of collaborative efforts from esteemed experts, each contributing a unique perspective to unveil the captivating interplay of chemicals in ecological systems. The journey commences with an enlightening exploration in Chapter 1, providing an introduction to the foundational concepts of chemical ecology, tracing its evolution and recognising its vital role in our understanding of ecological phenomena. As we navigate through the chapters, we delve into the captivating terrain of semiochemicals in Chapter 2, unravelling their significance, production and applications in pest management and beyond. Chapter 3 unveils the sensory world of insects, exploring chemoreception and shedding light on how insects navigate their surroundings and engage with the chemical cues that shape their behaviours. Chapter 4 unveils the complex mechanisms of chemical defence responses and their profound ecological implications. Diving deeper into the arsenal of plant defences, we decipher the art of host plant selection by herbivores in Chapter 5, uncovering the mechanisms that underpin this critical decision-making process. Our quest continues in Chapter 6, as we elevate our understanding of plant–herbivore interactions, from how plants detect pests to the intricate chemical conversations that define their relationships. Bridging this dynamic field with agriculture, Chapter 7 illuminates how chemical ecology shapes agricultural practices, pest management strategies and the sustainability of our food systems. Chapter 8 presents practical insights and methodologies instrumental in unravelling the mysteries of chemical ecology. This book invites readers to delve into the captivating realm of chemical ecology, where they can discover how molecules interact in intricate ways, much like dancers performing elaborate routines, contributing to the harmonious symphony of nature. Each chapter is a thread woven into the tapestry of ecological knowledge, illuminating the grandeur and complexity of the intricate relationships that bind organisms in our natural world.

About the Authors

Dr. Jamin Ali, an expert in applied insect chemical ecology, focuses on finding sustainable solutions for pest management worldwide. With a Ph.D. from Keele University, UK, and a master's from Aligarh Muslim University, India, he excelled in competitive exams such as JRF and GATE. Currently a foreign research expert at Jilin Agricultural University, China, Dr. Ali's research has taken him to countries including Kenya, the Netherlands, the United Kingdom, Israel, Malaysia and China. His work has been internationally recognised, earning him awards such as the National Overseas Scholarship and Türkiye Burslary Research Scholarship. Apart from his academic achievements, Dr. Ali actively shares his knowledge by publishing in journals and contributing to books. He engages with the scientific community through conference presentations and serves as a reviewer for international journals. Dr. Ali's practical approach and dedication make him a key contributor to sustainable pest management in agriculture globally.

Dr. Ri Zhao Chen, a Professor at Jilin Agricultural University, China, holds a doctorate in plant protection. With a robust academic background, including a Ph.D. from Jilin Agricultural University, Dr. Chen specialises in plant protection, sex pheromones and integrated pest management. His impactful research, funded by prestigious institutions, focuses on projects such as digital early warning systems for pests and diseases. Dr. Chen's work extends to practical applications, evident in his utility model patents for innovative insect monitoring and pest control devices. A prolific author, Dr. Chen has contributed extensively to SCI journals, emphasising insect behaviour and control mechanisms. His editorial roles in various books reflect his commitment to knowledge dissemination. Recognised for his outstanding achievements, Dr. Chen has received accolades for teaching quality, course construction and contributions to natural science. His dedication positions him as a key figure in plant protection and integrated pest management.

Acknowledgements

We extend our heartfelt gratitude to everyone who has supported and guided us throughout the creation of *Chemical Ecology: Insect-Plant Interactions*. We thank our colleagues and friends, past and present, for their unwavering support and encouragement during the book's development. Special thanks to Professor Li Yu (Jilin Agricultural University, China) for providing us support throughout the writing of this book. We are also grateful to Dr. Mohammad Mukarram, Dr. Mohd Umar Farukh, Dr. Adil Tonğa, Dr. Mogeda Abdelhafez, Babu Saddam, Shane Alam and Tabish Hashmi for their invaluable insights. Additionally, we appreciate the assistance of our lab students Ji Yunliang, Wu Haichao, Feng Xiao, Sohail Abbas, Huang Jingxuan, Bilal Ahmad, Aleena Alam and Qin Weibo. We are grateful for the scholarly guidance provided by Professor Ahmet Bayram (Dicle University, Türkiye), whose review of the book proposal greatly enriched the chapters. We extend special thanks to Owen Alice, Senior Editor at CRC Press Taylor & Francis, for guiding us through the publication process along with the efforts of Amelia Bashford and Shikha Garg. Furthermore, we express our sincere thanks to all colleagues who kindly provided permission for the use of figures in this book. Our heartfelt thanks go to our families for their unwavering love and encouragement, which has been the cornerstone of this endeavour. Special thanks to Mr. Shuaib Anwar for his continuous support and for listening to me for hours regarding this project and others. We also sincerely appreciate the facilities available at the "Conservation Tillage Pest and Disease Monitoring Base" of Jilin Agricultural University, which made the completion of this project possible.

Acronyms

ABA	Abscisic acid
AMF	Arbuscular mycorrhizal fungi
APS	Adenosine 5'-phosphosulphate
ATP	Adenosine triphosphate
AXR	Auxin resistant
BOA	Benzoxazolin-2-one
BRs	Brassinosteroids
C	Carbon
CEBiPs	Chitin elicitor binding proteins
CGs	Cyanogenic glycosides
CJ	*Cis*-Jasmone
CK	Cytokinins
CO2	Carbon dioxide
Cu	Copper
Cys	Cysteine
D	Dimension
DIBOA	2,4-dihydroxy-1,4-benzoxazin-3-one
DNA	Deoxyribonucleic acid
DWV	Deformed wing virus
EIN2	Ethylene insensitive2
ET	Ethylene
GAs	Gibberellins
GC-EAG	Gas chromatography-electroantennography
GC-MS	Gas chromatography-mass spectrometry
GC-SCR	Gas chromatography-single cell recordings
GRNs	Gustatory receptor neurons
GRs	Gustatory receptors
GS	Glucosinolates
HIPVs	Herbivore-induced plant volatiles
IAA	Indole-3-acetic acid

IBA	Indole-3-butyric acid
IPM	Integrated pest management
IPP	Isopentenyl diphosphate
IRs	Ionotropic receptors
JA	Jasmonic acid
JAs	Jasmonates
JAZ	JAZ genes
MAPP	Dimethylallyl diphosphate
MeJA	Methyl jasmonate
MEP	Methyl-D-erythritol-4-phosphate
MeSA	Methyl salicylate
Met	Methionine
MVA	Mevalonate
NPAAs	Non-protein amino acids
O3	Ozone
ORs	Odorant receptors
OSNs	Olfactory sensory neurons
PAPS	Phosphoadenosine 5'-phosphosulphate
PER	Proboscis extension responses
PGPR	Plant growth-promoting rhizobacteria
PIs	Proteinase inhibitors
POD	Peroxidase
PPO	Polyphenol oxidase
PR	Pathogenesis-related
PTFE	Polytetrafluoroethylene
ROS	Reactive oxygen species
SA	Salicylic acid
SAM	S-adenosylmethionine
SAR	Systemic acquired resistance
SAs	Salicylates
SDDS	Stimulo-deterrent diversionary strategies
Se	Selenium
SEZ	Suboesophageal zone
SO42-	Sulphate
SRs	Sugar receptors
TPI	As trypsin proteinase inhibitors
UV	Ultraviolet
VOCs	Volatile organic compounds
Zn	Zinc

Glossary

Term   **Description**

Allelochemicals   Chemicals released by one organism to affect the behaviour or physiology of other organisms.

Behavioural Bioassay   Tests designed to study insect behaviour in response to chemical cues.

Biological Control   The use of natural enemies to control pest populations.

Chemical Analysis   Laboratory methods for identifying and quantifying chemical compounds.

Chemical Communication   The exchange of information through the emission and detection of semiochemicals.

Chemical Communication   Exchange of chemical signals between plants and herbivores to influence behaviour.

Chemical Ecology   The study of chemical signals and their role in mediating interactions between organisms in natural environments.

Chemicals in Agriculture   The role of chemical compounds in crop production and protection.

Chemoreception   The ability of organisms to detect and respond to chemical stimuli.

Chemoreceptors   Sensory structures on insect antennae and mouthparts that detect chemical cues.

Coevolutionary Arms Race   Ongoing evolutionary competition between plants and herbivores.

Constitutive Defences   Pre-existing plant traits that deter herbivores.

Coping with Plant Defences   Strategies employed by herbivores to overcome plant chemical defences.

Electrical Penetration Graph (EPG)   Method for recording insect feeding behaviours.

Electrophysiology   Study of electrical signals generated by chemoreceptors in response to chemicals.

Entrainment Collection Technique to capture and analyse airborne volatiles emitted by plants.

Herbivore Feeding Adaptations Behavioural and physiological traits that enable insects to exploit plant resources.

Host Plant Choice Factors influencing the selection of specific host plants by herbivores.

Host Plant Recognition Mechanisms by which herbivores identify suitable host plants.

Host Plant Selection Process by which herbivores choose plants for feeding and oviposition.

Induced Defences Plant responses triggered by herbivore feeding or damage.

Induced Responses Defensive reactions triggered in plants by herbivore attack or other environmental factors.

Insect–Plant Interaction Relationships between insects and plants encompassing feeding, reproduction and defence.

Insect–Plant Interactions Complex relationships between insects and plants, often involving chemical cues.

Kairomones Chemical signals used by one species to exploit the behaviour of another species.

Non-Target Effects Unintended ecological impacts of pesticides on non-target organisms.

Olfaction The sense of smell in insects, allowing them to detect and respond to odour molecules.

Olfactory System The anatomical structures responsible for detecting and processing odours in insects.

Performance Bioassay Experimental setup to assess insect fitness and performance.

Pheromones Chemical signals released by organisms to communicate with members of the same species.

Pheromones in Pest Control Use of insect pheromones for pest monitoring and management.

Phytochemicals Any of various biologically active compounds found in plant.

Phytohormones Chemicals produced by plants that regulate their growth, development, reproductive processes, longevity and even dead.

Plant Defences Mechanisms that plants use to deter or reduce herbivore damage.

Plant Detection Mechanisms   Plant systems for sensing and responding to herbivore attack.

Plant–Pest Interactions   The effects of plant chemicals on pest behaviour and physiology.

Priming   The process by which plants enhance their defences in response to prior exposure to stressors.

Push-Pull Strategy   Pest management approach utilising attractive and repellent chemicals.

Semiochemicals   Chemical substances used for communication between organisms, including pheromones, allelochemicals and kairomones.

Signalling Pathways   Complex molecular networks that transmit chemical signals and regulate physiological responses.

Specialist   Insect species that consume a few related species

Sustainable Agriculture   The integration of chemical ecology principles for ecologically sound crop management.

CHAPTER 1

Introduction to Chemical Ecology

Jamin Ali and Ri Zhao Chen

1.1 INTRODUCTION

Chemical ecology stands at the intersection of various scientific fields, showcasing the complex relationships between organisms and their chemical surroundings (Eisner and Meinwald, 1995; Vet, 1999). Chemical ecology aims to understand the intricate connections between living organisms by studying how they communicate through chemical signals (Bergström, 2007). Chemical ecology brings together different areas of science, such as biology, chemistry, ecology and behaviour, to understand how chemical signals affect life on Earth. It goes beyond traditional science fields and involves biologists, chemists, ecologists and ethologists. This approach sees the importance of chemical signals in how organisms interact with each other, aiming for a complete understanding of the natural world (Blomquist and Bagnères, 2010; Meiners, 2015). Whether it is the scent molecules released by flowers to attract pollinators or the defensive compounds produced by plants to deter herbivores, chemical ecology delves into the nuances of these molecular dialogues that have been honed over millennia. Chemical ecology reveals the profound realisation that chemical compounds are not mere biochemical entities; they are the messengers of nature's secrets (Engelberth et al., 2004). In ecosystems, organisms have evolved an astonishing array of chemical signals to communicate a

DOI: 10.1201/9781003479857-1

1

myriad of messages, from warnings of danger to invitations for mating (Wilson, 1965; Leung, Beukeboom and Buellesbach, 2022). These chemical dialogues extend far beyond individual species, encompassing entire ecosystems where interactions ripple through trophic levels (Whitham *et al.*, 2006). By studying these chemical exchanges, chemical ecologists uncover the intricate relationships that govern the balance of life in our world, providing insights into the mechanisms that sustain the delicate fabric of biodiversity. In conclusion, chemical ecology casts light on the hidden conversations that shape ecological relationships (Vet, 1999; Cardé and Millar, 2004). It highlights the fact that every living being, from microorganisms to tall trees, is an active participant in a symphony of chemical communication that orchestrates the grand theatre of life (Dicke and Takken, 2006). As this and subsequent chapters explore further, the dimensions and significance of chemical ecology will continue to unfold, revealing the remarkable ways in which nature's molecules are the threads that bind living organisms together.

1.2 HISTORICAL CONTEXT AND EVOLUTION OF CHEMICAL ECOLOGY

The roots of chemical ecology trace back to the earliest observations of chemical interactions in the natural world (Agosta, 1992). Throughout history, indigenous cultures recognised the potent scents of plants, using them for medicinal, ritualistic and protective purposes, thus hinting at the complex language of chemical communication occurring between organisms (Miller, 1940; Agosta, 1992). An understanding that would gradually mature into the scientific discipline of chemical ecology, these early perceptions, while rudimentary, laid the foundation for recognising the profound role of chemical cues in ecological interactions. Thomas Eisner, a prominent German entomologist and ecologist, played a key role in advancing the field of chemical ecology and is known as the "father of chemical ecology" (Berenbaum, 2012). This branch of ecology, focused on chemical interactions among organisms, emerged in the 1960s when E.O. Wilson and Robert MacArthur introduced the concept of chemical signals as a significant aspect of ecological study (Wilson, 1971; Moreau and Pierce, 2022).

As mentioned earlier, chemical ecology is a multidisciplinary field that involves the study of interactions between organisms and their chemical environments (Bergström, 2007). It encompasses the chemical compounds that organisms produce, their mechanisms of action and the ecological roles they play in the environment. The evolutionary significance of chemical adaptations became evident as researchers explored the diversity of life on Earth. Organisms had not only developed morphological and physiological adaptations but had also evolved intricate chemical strategies to survive and thrive (Bock and Von Wahlert, 1965). Chemical adaptations played a pivotal role in shaping coevolutionary dynamics between species (Lankau, 2012). The arms race between plants and herbivores exemplified the selective pressure driving the development of novel chemical defences and the counter-adaptations of herbivores to tolerate or detoxify these compounds (Strauss and Zangerl, 2002; Arora, 2012). This dance of coevolution and adaptation, mediated through chemicals, showcased the power of chemical ecology in driving the intricacies of ecological relationships and provided a compelling context for the development of this scientific discipline (Feeny, 1975; Merillon and Ramawat, 2020).

The field of chemical ecology is a relatively new discipline that emerged in the 1950s (Price and Hunter, 1995; Hartmann, 2008). Even though scientists have been aware for centuries that organisms can utilise chemicals for communication. Early studies in chemical ecology focused on the discovery of novel chemical compounds produced by organisms (Dyer et al., 2018). However, over the past few decades, the focus has shifted toward studying the ecological significance of these compounds (Mbaluto et al., 2020). The establishment of analytical techniques and instruments, such as gas chromatography and mass spectrometry, enabled scientists to delve deeper into the chemical compositions of plants, insects and their environments (Dyer et al., 2018). This newfound ability to decipher the intricate chemical dialogues sparked the birth of chemical ecology as a recognised field of study. Researchers, captivated by the role of chemical cues in guiding behaviours, unravelling species interactions and shaping ecosystems, started to collaborate across disciplines, uniting biology, chemistry and even physics to comprehend the complex relationships encoded in chemical compounds.

1.3 WHAT IS CHEMICAL ECOLOGY?

Chemical ecology is the study of chemical interactions between organisms and their environments. It is aimed at understanding how organisms utilise chemicals to communicate, defend themselves, find mates and obtain food (Wajnberg and Colazza, 2013; Kansman et al., 2023). This field combines aspects of biology, chemistry and ecology to investigate the complex chemical interactions that occur in natural systems (Bergström, 2007;

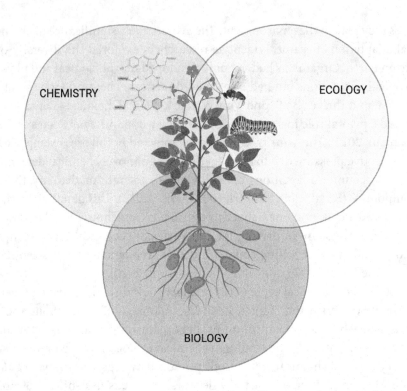

FIGURE 1.1   Illustration showing the main disciplines contributing to chemical ecology.

Kansman *et al.*, 2023; Vet, 1999) (Figure 1.1). The interactions investigated by chemical ecologists range from intra-specific interactions to those occurring between different species. This interdisciplinary field hopes to provide insights into natural phenomena and develop strategies for pest control, conservation biology and sustainable agriculture (Cumming and Spiesman, 2006; Jacquet *et al.*, 2022).

Here is how each of these disciplines contribute to chemical ecology:

- *Chemistry*: Chemical ecology relies heavily on the principles of organic chemistry to identify, isolate and quantify the chemical compounds that mediate ecological interactions. Chemists use various analytical techniques such as chromatography, mass spectrometry and nuclear magnetic resonance spectroscopy to analyse chemical compounds in different organisms and their environments.

- *Biology*: Biological principles are important to understand the roles of chemical signals in the behaviour, development and communication of different organisms. Biologists study the ecology and behaviour of different species to identify the ecological functions of chemical signals produced by different organisms.

- *Ecology*: Ecological principles help to understand the interactions between organisms and their environments, including their roles in food webs, predator–prey interactions and competition. Ecologists study different ecosystems to identify the chemical signals that mediate these interactions and how they are affected by biotic and abiotic factors.

Together, chemistry, biology and ecology are all essential for understanding how chemical signals play a crucial role in the survival, adaptation and evolution of different organisms in their respective ecosystems and how changes in chemical environments can have significant impacts on these interactions.

1.4 WHY IS CHEMICAL ECOLOGY IMPORTANT?

Chemical ecology is an interdisciplinary field that looks at the role of chemistry in interactions between organisms in their environments. It helps us to understand the complex food webs, communication systems and behaviours of organisms (Pasteels, Grégoire and Rowell-Rahier, 1983). Chemical ecology has practical applications in agriculture, medicine and pest control (Mbaluto *et al.*, 2020). Through chemical ecology research, we can develop environmentally friendly methods of controlling pests and diseases that do not harm beneficial organisms (Witzgall, Kirsch and Cork, 2010), as some chemicals from plants can be used in the development of new drugs. Chemical signals play a vital role in the communication between different species, including attractants, repellents and mating pheromones (Witzgall, Kirsch and Cork, 2010). These compounds can also signal to predators that they should avoid preying on a specific plant species (Halitschke *et al.*, 2008). Understanding this communication is critical for developing effective methods for pest control, disease management and conservation.

Chemical ecology deals with the interactions between organisms and their environments, as mediated by the chemical signals they produce and receive. It extends from the molecular level to the ecosystem level and involves the study of the biosynthesis, perception and ecological roles of diverse natural products, including pheromones, allelochemicals and secondary metabolites (Dicke and Takken, 2006). At the molecular level, chemical ecology involves the genetic and biochemical mechanisms that govern the synthesis and regulation of chemical signals and their receptors (Jacquin-Joly and Merlin, 2004). At the organismal level, it involves the study of how chemical signals mediate social behaviour, competition and prey–predator interactions. At the ecosystem level, chemical ecology involves the study of how these interactions shape community structure, nutrient cycling and ecosystem functions (Dicke and Takken, 2006). By integrating across these levels of biological organisation, chemical ecology provides a holistic approach to understanding the links between genes, individuals, communities and ecosystems.

The principle of chemical ecology is the study of the chemical interactions between living organisms and their environments. This can include the production and reception of chemical signals used in communication, the use of chemical defences against predators and competitors and the chemical modification of physical and biological environments (Abbot, Tooker and Lawson, 2018; Dyer *et al.*, 2018). Chemical ecology involves the study of the role of chemicals in ecosystem function, the evolution of these chemical interactions and the potential for using these chemicals in applied settings, such as in agriculture or medicine. This multidisciplinary approach not only deepens our understanding of ecological processes but also holds the promise of practical applications across diverse domains.

- *Understanding the interactions between organisms:* Chemical signals play a critical role in the communication and interactions between various organisms, from bacteria and plants to insects and mammals. By studying these signals, researchers can gain a better understanding of the relationships between different species and how they coexist in their environment. Moreover, this knowledge has far-reaching implications.

- *Developing new pest control strategies:* Herbivorous insects use chemical signals to communicate and coordinate their behaviour. Understanding these signals can help researchers to develop new pest control strategies that disrupt their communication and reduce their damage to crops and other ecosystems. This innovative approach to pest management holds great potential for sustainable agricultural practices.

- *Discovering new pharmaceuticals:* Many organisms, particularly plants, produce chemicals that have medicinal properties. By studying the chemical signals produced by these organisms, researchers can discover new compounds that may have potential as drugs to treat a variety of human diseases. This intersection of chemistry and biology opens avenues for novel medical breakthroughs.

- *Conserving endangered species:* Chemical signals are often critical for reproduction and other important behaviours in many species. By studying these signals, researchers can develop strategies to protect and conserve endangered species by improving their habitat and supporting their reproduction and survival. This interdisciplinary pursuit contributes significantly to biodiversity preservation and ecosystem health.

1.5 CHEMICAL COMMUNICATION IN AN ECOLOGICAL CONTEXT

Chemical cues are essential components that intricately weave the fabric of ecological interactions, particularly within the context of insect–plant relationships (Nishida, 2014). Organisms rely on chemical signals to communicate a variety of crucial information, transcending language barriers to convey vital messages in their environments (Chagas *et al.*, 2018; Zu *et al.*, 2023). In the complex interplay between insects and plants, chemical cues play a pivotal role in shaping their interactions and influencing their behaviours (Zu *et al.*, 2023). One of the key roles of chemical signals is species recognition (Wyatt, 2014). Insects, often existing in highly diverse communities, need effective mechanisms to differentiate between conspecifics and other species (Johansson and Jones, 2007). Chemical cues emitted by insects and plants serve as distinct "chemical signatures," enabling insects to identify their own species and, equally importantly, recognise potential mates (Johansson and Jones, 2007). This recognition is fundamental for successful reproduction, as compatible mates are located based on these chemical cues, ensuring the genetic diversity and stability of populations.

Moreover, chemical signals are not solely confined to matters of courtship. They extend their influence into the intricate web of predator–prey relationships (Zu *et al.*, 2023). For instance, plants may release specific volatile compounds when attacked by herbivores. These compounds act as a distress signal, alerting predators to the presence of prey (Baldwin, 2010).

Predatory insects can then home in on these chemical cues to locate their next meal (Hatano *et al.*, 2008). This process creates a chain of interactions in which chemical cues act as the messengers facilitating the orchestration of the ecological symphony (Rasmussen and Schulte, 1998).

1.6 APPLICATIONS OF CHEMICAL ECOLOGY

Chemical ecology finds wide-ranging applications across various domains, including agriculture, pest management and pharmaceuticals. In agriculture, the insights garnered from studying chemical interactions between plants, insects and other organisms have revolutionised crop protection strategies (Mbaluto *et al.*, 2020). By deciphering the chemical signals that attract or repel herbivores, researchers can develop targeted pest control methods (Kansman *et al.*, 2023). For instance, the identification and synthesis of pheromones, which are specific chemicals released by insects to communicate with others of the same species, have paved the way for effective monitoring and trapping of pest populations (El-Shafie and Faleiro, 2017). Additionally, the integration of chemical cues into crop breeding programs has led to the development of pest-resistant plant varieties, reducing the reliance on chemical pesticides and promoting sustainable agricultural practices (Kansman *et al.*, 2023).

Chemical ecology also contributes significantly to our understanding of ecosystem dynamics and biodiversity. The intricate web of chemical interactions shapes the structure and functioning of ecosystems (Cortesero *et al.*, 2016). By investigating how organisms use chemical cues for foraging, mating and communication, scientists gain insights into trophic interactions and the interconnectedness of species within food webs (Raguso *et al.*, 2015). These revelations help elucidate the role of specific species in maintaining ecosystem balance and functioning. Furthermore, chemical cues play a pivotal role in mediating predator–prey relationships, leading to the discovery of natural pest control mechanisms (Hatano *et al.*, 2008; Zu *et al.*, 2023). Harnessing these insights, researchers have developed innovative approaches that capitalise on the natural enemies of pests to regulate populations, reducing the need for synthetic chemicals and minimising ecological disruptions.

In the pharmaceuticals industry, chemical ecology holds promise for drug discovery and development (Wolfender *et al.*, 2019). Many organisms, particularly plants and marine organisms, produce bioactive compounds with potential medicinal properties. By studying the chemical interactions and defensive mechanisms of these organisms, researchers can identify compounds that could have therapeutic applications (Dillard and German, 2000). Moreover, understanding the chemical ecology of disease vectors, such as mosquitoes, can lead to the development of novel methods for disease control (Mwingira *et al.*, 2020). The field of chemical ecology, thus, offers a treasure trove of opportunities to unlock nature's chemical arsenal for the betterment of agriculture, human health and ecosystem sustainability.

1.7 CHEMICAL ECOLOGY AND INSECT–PLANT INTERACTION

Plants and insects have coexisted for millions of years, resulting in a variety of interactions ranging from beneficial ones, such as pollination and seed dispersion, to detrimental ones caused by herbivorous insects. Insect herbivores primarily rely on plants to fulfil various purposes of life, such as food, shelter, oviposition. Similarly, in response to herbivorous insects, plants have developed a wide range of defence strategies. Among these, chemical defences possess unique modes of action, including both direct and indirect approaches to protect themselves. These interactions between insect and plant are extensively governed by the involvement of chemicals. Recognition of plant chemicals by herbivorous insects help them to detect host and non-host plants. Similarly, insect feeding and oviposition are associated with some kind of chemical secreted by herbivorous insects. These chemicals are generally known as semiochemicals. In chemical ecology, semiochemicals emerge as crucial elements facilitating communication between diverse organisms. This communication significantly influences the intricate relationships within ecosystems, particularly in the context of insect–plant interactions. Pheromones, species-specific chemical signals crafted by insects, serve as potent communicators within their communities. They relay information about mating, territory marking and alarm signalling and are finely tuned to elicit specific responses from individuals of the same species, thereby profoundly influencing their behaviour and physiology. Allelochemicals, synthesised by plants, constitute another essential component in the intricate dance between flora and fauna. These chemical compounds possess the remarkable ability to influence the behaviour or physiology of other organisms, often acting as a form of defence against herbivores. The diverse array of allelochemicals showcases the adaptability and sophistication of plant defence mechanisms. Kairomones introduce an additional layer of complexity, representing chemical signals released by one species to benefit another. In the context of insect–plant interactions, certain plants release kairomones to attract the attention of natural enemies of herbivores. This indirect defence strategy leverages ecological relationships, utilising kairomones to summon predators or parasitoids that contribute to the regulation of herbivore populations. Semiochemicals, as dynamic mediators of communication, transcend the boundaries of individual organisms, weaving a tapestry of interactions that extends across species. Chapter 2 is particularly dedicated to semiochemicals.

Phytochemicals have been extensively used in implementation of chemical ecology. Understanding the collection and identification of phytochemicals alongside the intricacies of insect chemoreception is pivotal for various disciplines such as agriculture and ecology. It involves delving into the mechanisms of insect olfaction, including the detailed understanding of how insects detect and respond to chemical stimuli. This exploration encompasses the examination of insect chemoreceptors, their unique expression features and the adaptive evolution of chemoreceptive structures for enhanced detection and perception of phytochemicals. Detailed information about chemoreception in insects is discussed in Chapter 3, providing further insights into this fundamental aspect of insect chemoreception.

Furthermore, chemical ecology plays a pivotal role in plant defence against herbivorous insects by employing phytochemicals as crucial components of this defence mechanism. These chemicals deter herbivores by producing toxic compounds that render plant tissues unpalatable or harmful upon ingestion, effectively reducing herbivory rates and minimising damage to the plant. Additionally, phytochemicals impede the growth and development of insect herbivores by disrupting physiological processes or inhibiting crucial biochemical pathways within the herbivore's body, thereby hindering their ability to thrive and reproduce. Furthermore, these chemicals facilitate the recruitment of biocontrol agents, such as predators and parasitoids, by emitting cues that attract beneficial organisms to the plant. Through this indirect defence mechanism, plants bolster their protection against herbivores by promoting the suppression of pest populations. Detailed discussions on this aspect of chemical ecology in the context of insect–plant interactions are provided in Chapter 4, highlighting the intricate interplay between phytochemicals, herbivorous insects and natural enemies and elucidating the mechanisms by which plants deploy chemical defences to enhance their survival.

The interaction between insects and host plants is profoundly influenced by host plant chemistry, which shapes various aspects of this intricate relationship. First, the behaviour and performance of insects are directly and indirectly linked to the status of host plants. This status encompasses the physiological condition, nutritional value and chemical composition of the host plant. Herbivorous insects assess the status of potential host plants before considering them suitable for feeding and oviposition. Chemical cues emitted by host plants play a pivotal role in this selection process, guiding insects toward suitable hosts and influencing their feeding preferences. Moreover, the phenomenon of host plant specialisation versus generalisation among herbivorous insects reflects the intricate interplay between host plant chemistry and insect biology. Some insect species exhibit specialisation toward specific host plants, while others display a broader dietary range, depending on the chemical cues emitted by different host plants. The quality of the host plant also exerts a significant impact on insect performance, influencing factors such as growth, development and reproduction. Host plants with optimal nutritional quality and chemical composition promote better performance in associated herbivorous insects, while inferior hosts may hinder insect fitness. Understanding the implications of host plant selection for pest management is crucial in devising effective strategies for agricultural and ecological applications. Chapter 5 delves into this aspect in detail, elucidating the intricate relationships between host plant chemistry, insect behaviour and the implications for pest management strategies.

Another important aspect of the chemical ecology of insect–plant interaction is its role in understanding how it shapes the dynamics between these organisms and influences their performance. From a general perspective, chemical cues exchanged between plants and insects not only serve as signals for recognition but also impact their population dynamics and adaptations. Upon recognising chemical cues released by host plants, insects can inflict significant damage on plants through direct feeding and indirect damage, such as spreading plant viruses that lead to diseases. To elucidate the nature of plant adaptations in response to different insect feeding guilds, we focus on *brassica* plants. We explore the diverse defence responses employed by plants and how these defences vary in response to different feeding guilds, including sucking and chewing insects. Additionally, we delve into how insects adapt to these defence responses and the role of chemical signalling underlying these interactions. These discussions are extensively covered in Chapter 6, shedding light on the intricate interplay between chemical cues, plant defences, insect adaptations and their collective impact on the dynamics of insect–plant interactions.

Continuing our exploration, Chapter 7 delves into the pivotal role of chemical ecology within agricultural contexts, showcasing its contribution to the development of pest management strategies, such as push-pull, seed priming and the implementation of lures and traps. Moving forward to Chapter 8, we delve into fundamental techniques essential to chemical ecology. Here, we detail methods such as performance and behavioural bioassays, including olfactometer studies, as well as procedures for plant collection and identification employing gas chromatography (GC-MS) and gas chromatography electroantennography (GC-EAG) methodologies.

1.8 CONCLUSION

In conclusion, this chapter has provided a comprehensive overview of chemical ecology, beginning with an introduction to the field and tracing its historical context and evolution. We have explored the fundamental concepts of chemical ecology, emphasising its significance in understanding ecological interactions. From chemical communication to its diverse applications, the role of chemical ecology in elucidating insect–plant interactions has been highlighted. Through this exploration, it becomes evident that chemical ecology offers invaluable insights into the complexities of ecological systems and plays a crucial role in informing conservation efforts and sustainable practices.

REFERENCES

Abbot, P., Tooker, J. and Lawson, S. P. (2018) 'Chemical ecology and sociality in aphids: opportunities and directions', *Journal of Chemical Ecology*, 44(9), pp. 770–784. doi: 10.1007/s10886-018-0955-z.

Agosta, W. C. (1992) *Chemical communication: the language of pheromones.* Henry Holt and Company.

Arora, R. (2012) *Co-evolution of insects and plants.* Scientific Publications Jodhpur.

Baldwin, I. T. (2010) 'Plant volatiles', *Current Biology*, 20(9), pp. 392–397. doi: 10.1016/j.cub.2010.02.052.

Berenbaum, M. (2012) 'Thomas Eisner 1929–2012', *Bulletin of the Ecological Society of America*, 93(3), pp. 191–196.

Bergström, G. (2007) 'Chemical ecology= chemistry+ ecology!', *Pure and Applied Chemistry*, 79(12), pp. 2305–2323.

Blomquist, G. J. and Bagnères, A.-G. (2010) *Insect hydrocarbons: biology, biochemistry, and chemical ecology.* Cambridge University Press.

Bock, W. J. and Von Wahlert, G. (1965) 'Adaptation and the form-function complex', *Evolution*, 19, pp. 269–299.

Cardé, R. T. and Millar, J. G. (2004) *Advances in insect chemical ecology.* Cambridge University Press.

Chagas, F. O. *et al.* (2018) 'Chemical signaling involved in plant–microbe interactions', *Chemical Society Reviews*, 47(5), pp. 1652–1704.

Cortesero, A. *et al.* (2016) 'Chemical ecology: an integrative and experimental science', *Chemical ecology*, pp. 23–46.

Cumming, G. S. and Spiesman, B. J. (2006) 'Regional problems need integrated solutions: pest management and conservation biology in agroecosystems', *Biological Conservation*, 131(4), pp. 533–543.

Dicke, M. and Takken, W. (2006) *Chemical ecology: from gene to ecosystem.* Springer Science & Business Media.

Dillard, C. J. and German, J. B. (2000) 'Phytochemicals: nutraceuticals and human health', *Journal of the Science of Food and Agriculture*, 80(12), pp. 1744–1756.

Dyer, L. A. *et al.* (2018) 'Modern approaches to study plant–insect interactions in chemical ecology', *Nature Reviews Chemistry*, 2(6), pp. 50–64.

Eisner, T. and Meinwald, J. (1995) 'Chemical ecology', *Proceedings of the National Academy of Sciences*, 92(1), p. 1.

El-Shafie, H. A. F. and Faleiro, J. R. (2017) 'Semiochemicals and their potential use in pest management', *Biological control of pest and vector insects*, pp. 10–5772.

Engelberth, J. *et al.* (2004) 'Airborne signals prime plants against insect herbivore attack', *Proceedings of the National Academy of Sciences of the United States of America*, 101(6), pp. 1781–1785. doi: 10.1073/pnas.0308037100.

Feeny, P. (1975) 'Biochemical coevolution between plants and their insect herbivores', *Coevolution of animals and plants: symposium V, first international congress of systematic and evolutionary biology, 1973.* University of Texas Press, pp. 1–19.

Halitschke, R. *et al.* (2008) 'Shared signals–"alarm calls" from plants increase apparency to herbivores and their enemies in nature', *Ecology Letters*, 11(1), pp. 24–34.

Hartmann, T. (2008) 'The lost origin of chemical ecology in the late 19th century', *Proceedings of the National Academy of Sciences*, 105(12), pp. 4541–4546.

Hatano, E. *et al.* (2008) 'Chemical cues mediating aphid location by natural enemies', *European Journal of Entomology*, 105(5), pp. 797–806.

Jacquet, F. *et al.* (2022) 'Pesticide-free agriculture as a new paradigm for research', *Agronomy for Sustainable Development*, 42(1), p. 8.

Jacquin-Joly, E. and Merlin, C. (2004) 'Insect olfactory receptors: contributions of molecular biology to chemical ecology', *Journal of Chemical Ecology*, 30, pp. 2359–2397.

Johansson, B. G. and Jones, T. M. (2007) 'The role of chemical communication in mate choice', *Biological Reviews*, 82(2), pp. 265–289.

Kansman, J. T. *et al.* (2023) 'Chemical ecology in conservation biocontrol: new perspectives for plant protection', *Trends in Plant Science*, 28, pp. 1166–1177.

Lankau, R. A. (2012) 'Coevolution between invasive and native plants driven by chemical competition and soil biota', *Proceedings of the National Academy of Sciences*, 109(28), pp. 11240–11245.

Leung, K., Beukeboom, L. W. and Buellesbach, J. (2022) 'Novel insights into insect chemical communication–an introduction', *Entomologia Experimentalis et Applicata*. Wiley Online Library, pp. 286–288.

Mbaluto, C. M. *et al.* (2020) 'Insect chemical ecology: chemically mediated interactions and novel applications in agriculture', *Arthropod-plant Interactions*, 14, pp. 671–684.

Meiners, T. (2015) 'Chemical ecology and evolution of plant–insect interactions: a multitrophic perspective', *Current Opinion in Insect Science*, 8, pp. 22–28.

Merillon, J.-M. and Ramawat, K. G. (2020) *Co-evolution of secondary metabolites*. Springer.

Miller, E. M. (1940) 'Chemical intergrative mechanisms in insect societies', *Proceedings of the Florida Academy of sciences*. JSTOR, pp. 136–147.

Moreau, C. S. and Pierce, N. E. (2022) 'Edward O. Wilson (1929–2021)', *Nature Ecology & Evolution*, 6(3), pp. 240–241.

Mwingira, V. *et al.* (2020) 'Exploiting the chemical ecology of mosquito oviposition behavior in mosquito surveillance and control: a review', *Journal of Vector Ecology*, 45(2), pp. 155–179.

Nishida, R. (2014) 'Chemical ecology of insect–plant interactions: ecological significance of plant secondary metabolites', *Bioscience, Biotechnology, and Biochemistry*, 78(1), pp. 1–13.

Pasteels, J. M., Grégoire, J.-C. and Rowell-Rahier, M. (1983) 'The chemical ecology of defense in arthropods', *Annual Review of Entomology*, 28(1), pp. 263–289.

Price, P. W. and Hunter, M. D. (1995) 'Novelty and synthesis in the development of population dynamics', *Population dynamics: new approaches and synthesis*, pp. 389–412.

Raguso, R. A. *et al.* (2015) 'The raison d'être of chemical ecology', *Ecology*, 96(3), pp. 617–630.

Rasmussen, L. E. L. and Schulte, B. A. (1998) 'Chemical signals in the reproduction of Asian (Elephas maximus) and African (Loxodonta africana) elephants', *Animal Reproduction Science*, 53(1–4), pp. 19–34.

Strauss, S. Y. and Zangerl, A. R. (2002) 'Plant-insect interactions in terrestrial ecosystems', *Plant-animal interactions: an evolutionary approach*, 2002, pp. 77–106.

Vet, L. E. M. (1999) 'From chemical to population ecology: infochemical use in an evolutionary context', *Journal of Chemical Ecology*, 25, pp. 31–49.

Wajnberg, E. and Colazza, S. (2013) *Chemical ecology of insect parasitoids*. John Wiley & Sons.

Whitham, T. G. *et al.* (2006) 'A framework for community and ecosystem genetics: from genes to ecosystems', *Nature Reviews Genetics*, 7(7), pp. 510–523.

Wilson, E. O. (1965) 'Chemical communication in the social insects: Insect societies are organized principally by complex systems of chemical signals', *Science*, 149(3688), pp. 1064–1071.

Wilson, E. O. (1971) 'Chemical communication within animal species', *Chemical ecology*, pp. 133–135.

Witzgall, P., Kirsch, P. and Cork, A. (2010) 'Sex pheromones and their impact on pest management', *Journal of Chemical Ecology*, 36(1), pp. 80–100.

Wolfender, J.-L. *et al.* (2019) 'Innovative omics-based approaches for prioritisation and targeted isolation of natural products–new strategies for drug discovery', *Natural Product Reports*, 36(6), pp. 855–868.

Wyatt, T. D. (2014) *Pheromones and animal behavior: chemical signals and signatures.* Cambridge University Press.

Zu, P. *et al.* (2023) 'Plant–insect chemical communication in ecological communities: an information theory perspective', *Journal of Systematics and Evolution*, 61(3), pp. 445–453.

CHAPTER 2

Semiochemicals

2.1 INTRODUCTION

In the complex network of ecological interactions, in which organisms engage in a complex symphony of communication, adaptation and survival, semiochemicals emerge as the delicate threads that weave together the stories of countless species. The term "semiochemicals" encompasses a diverse array of chemical compounds that serve as messengers in the intricate dialogues between organisms (Dicke and Sabelis, 1988). The term "semiochemicals," introduced by Whittaker in 1970, refers to chemical compounds that facilitate interactions between organisms of different species. These chemical signals, whether intentionally emitted or unconsciously released, play a pivotal role in shaping behaviours, mediating interactions and influencing the very fabric of ecosystems (Bruce and Pickett, 2011; Karban, 2015). As we embark on the exploration of semiochemicals in this book, we delve into a realm where molecules carry messages, shape behaviours, and drive the exquisite choreography of life.

Semiochemicals, also referred to as "infochemicals," are the chemical messengers that facilitate communication and interaction between organisms (Dicke and Sabelis, 1988). This encompasses a broad spectrum of compounds, from volatile aromatics that waft through the air, guiding insects to suitable mates or oviposition sites, to complex blends of pheromones that relay intricate information about species identity, readiness to mate or alarm cues (Karban, 2015). Beyond their fundamental roles in communication, semiochemicals serve as agents of influence, steering behaviours that impact ecological dynamics in profound ways. From deterring

DOI: 10.1201/9781003479857-2

herbivores from feeding on certain plants to luring predators toward prey, semiochemicals embody the intricate language of life (Cook *et al.*, 2007). This chapter aims to explore the multifaceted world of semiochemicals, uncovering their mechanisms, diversity and ecological implications. It commences by elucidating the types of semiochemicals, examining the distinction between allelochemicals, kairomones, synomones and other classes that are pivotal in shaping ecological interactions. The investigation into semiochemicals leads us into the realm of chemical signalling, in which the intricacies of how organisms perceive and respond to these chemical cues are explored. This entails analysing the sensory mechanisms and receptors that have evolved to decode the messages embedded within semiochemicals.

2.2 CLASSIFICATION

Semiochemicals can be categorized into two types, namely pheromones, which facilitate communication among individuals of the same species, and allelochemicals, which are compounds participating in communication between individuals of distinct species, based on the identity of the emitter and the receiver (El-Ghany, 2019). Allelochemicals can be further classified into various types based on their roles and functions within the intricate web of life (Dicke and Sabelis, 1988; Komala, Manda and Seram, 2021). As we explore the diversity of semiochemicals, we uncover a network of communication strategies that have evolved across species and ecosystems. Some semiochemicals are beneficial for the emitter, while some have advantages for the receiver (Scolari *et al.*, 2021) (Figure 2.1).

2.2.1 Allelochemicals

Allelochemicals play a pivotal role in shaping the intricate dynamics between insects and plants, exerting a profound influence on both ecological systems and sustainable agriculture (Macias *et al.*, 2007). These secondary metabolites, produced by various plant species, serve as potent chemical weapons, influencing herbivores' feeding behaviour, growth and survival (Barbosa and Letourneau, 1988; Divekar *et al.*, 2022). By acting as natural repellents or toxins, allelochemicals act as a plant's first line of defence against herbivorous insects, reducing the damage caused by pests (Qi *et al.*, 2020; Tlak Gajger and Dar, 2021). Moreover, allelochemicals can also function as attractants for beneficial insects, such as pollinators and natural predators, fostering a balanced ecosystem that enhances agricultural productivity (Tlak Gajger and Dar, 2021). Harnessing the power of

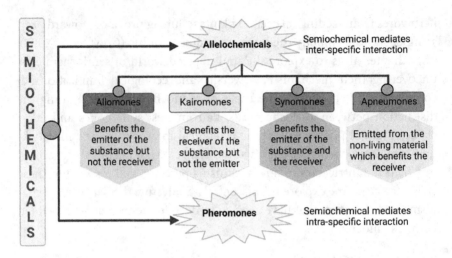

FIGURE 2.1 A brief classification of semiochemicals with definition.

allelochemicals in integrated pest management strategies can contribute to reducing the reliance on synthetic pesticides, promoting sustainable agricultural practices that protect both crop yields and environmental health (Regnault-Roger and Philogène, 2008; Hickman *et al.,* 2021; Tlak Gajger and Dar, 2021). In this way, understanding and leveraging the significance of allelochemicals in insect–plant interactions is a critical component of fostering resilient and eco-friendly agricultural systems. As mentioned earlier, allelochemicals can be further classified into four categories: allomones, kairomones, synomones and apneumones.

- *Allomones:* Allomones, derived from the Greek words *allos* (meaning "others") and *hormon* (meaning "excite"), are substances released by one organism that trigger a response in another species, typically benefiting the emitter. Examples of allomones include the defensive secretions produced by certain insects, which are toxic or repellent to their predators. For instance, some butterfly larvae possess cervical glands that release defensive secretions when disturbed. Allomones also serve as defence mechanisms for plants, deterring herbivores and reducing competition with neighbouring plants. Additionally, some social wasps and even certain flowers have evolved allomones to repel unwanted visitors or attract specific pollinators, such as bats. Interestingly, plants also produce "secondary products" as allomones – substances seemingly unrelated to their physiology but crucial for defence against herbivores and other animals.

- *Kairomones:* Kairomones are chemical compounds emitted by one organism, typically a plant, unintentionally benefiting another organism, often an herbivorous insect (Brown Jr, Eisner and Whittaker, 1970). These signals serve as inadvertent cues that attract herbivores or natural enemies of herbivores, indirectly aiding the receiver (Sbarbati and Osculati, 2006). For example, a plant may release kairomones that attract predators, helping to control herbivore populations.

- *Synomones:* In contrast, synomones are chemical compounds actively produced by an organism, usually a plant, for mutual benefit (Meiners and Hilker, 2000). In the context of insect–plant interactions, plants release synomones when under herbivore attack or oviposition, signalling to beneficial organisms, such as parasitoids or predators, that herbivores are present (Salerno *et al.*, 2013). In this case, both the emitter (plant) and the receiver (beneficial organism) benefit from this chemical signalling.

- *Apneumones:* Apneumones are distinct in that they are emitted from non-living materials, such as decaying organic matter or insect frass (excrement), to benefit a receiver, often by providing cues or resources (Hansson and Wicher, 2016; Bakthatvatsalam, Subharan and Mani, 2022). For instance, the scent of decaying plant matter can attract detritivores that aid in breaking down organic material (LeBlanc, 2021).

The importance of these chemical signals in insect–plant interactions and sustainable agriculture cannot be overstated. Understanding kairomones and synomones allows us to develop innovative pest management strategies that rely less on synthetic pesticides (Blassioli-Moraes *et al.*, 2019). By harnessing the power of these signals, we can promote natural pest control and reduce the ecological impact of agriculture while simultaneously safeguarding crop yields (Tlak Gajger and Dar, 2021). Apneumones, although less studied, may also hold potential for ecological applications in agriculture, further enhancing sustainability efforts. Individual volatile compounds may belong to more than one of the above groups. The same molecule can act as a pheromone for one species and as an allelochemical for another species. For example, aphids release (E)-β-farnesene under the attack of parasitoids. This compound elicits a repellent response and disperses nearby aphids (Nault, Edwards and Styer, 1973; Dicke and Sabelis, 1988). In this situation (E)-β-farnesene acts as a pheromone because it is used for intra-specific communication. However, when a compound exhibits the same function across different species of aphids, it is called an allelochemical. For instance, (E)-β-farnesene is an allelochemical that attracts natural enemies to caterpillar-damaged plants (Schnee *et al.*, 2006). Plants produce a plethora of volatile compounds that allow quick defence signalling between distant plant organs (Heil and Bueno, 2007), communication between plants (Baldwin *et al.*, 2006) and recruitment of natural enemies (Kessler and Baldwin, 2001). Plant volatiles may be released because of plant tissue damage or induction by an elicitor upon insect feeding. Phytophagous insects damage plant tissue with their mouthparts, secrete elicitors in their saliva and cause the release of herbivore-induced plant volatiles (De Moraes *et al.*, 1998; Paré and Tumlinson, 1999; Engelberth *et al.*, 2004; Mithöfer, Boland and Maffei, 2009).

Allelochemicals constitute a fascinating form of semiochemicals within the plant kingdom (Regnault-Roger and Philogène, 2008; Shorey, 2013). Allelochemicals are chemical compounds released by plants that influence the growth, development, and interactions of other plant species in their vicinity (Regnault-Roger and Philogène, 2008). These compounds act as chemical messages, conveying information about the sending plant's physiological state and readiness to compete for resources (Dicke *et al.*, 1990).

An intriguing example of allelochemicals is the phenomenon of allelopathy, in which certain plants release chemicals that inhibit the growth of neighbouring plants. The black walnut tree (*Juglans nigra*) secretes juglone, a compound that hampers the growth of many other plant species around it (Jilani *et al.*, 2008). This chemical strategy provides the black walnut with a competitive advantage, enabling it to dominate its immediate environment by suppressing the growth of potential competitors.

2.2.2 Pheromones

Pheromones stand as one of the most iconic examples of semiochemicals, playing a pivotal role in communication within species (Christensen and Sorensen, 1996; Komala, Manda and Seram, 2021). These chemical compounds are produced by organisms and elicit specific behavioural responses in conspecific individuals of the same species (Abd El-Ghany, 2020). Pheromones have been observed in a myriad of organisms, ranging from insects and mammals to even certain plant species (Dulac and Torello, 2003; Yew and Chung, 2015).

Sex pheromones are perhaps the most well-studied type of pheromones, and they orchestrate the intricate dance of reproduction. In many insect species, female moths emit sex pheromones that attract males from considerable distances (Jacobson, 2012). These pheromones are often composed of complex blends of volatile compounds, each conveying subtle information about the female's age, reproductive status and species identity (Jacobson, 2012; Yew and Chung, 2015). Male insects, equipped with specialised antennae designed to detect even trace amounts of these compounds, follow these chemical trails with astonishing precision (Jacobson, 2012).

Aggregation pheromones provide another compelling example of pheromone communication (Shorey, 1973). Insects such as bark beetles release aggregation pheromones to signal the presence of suitable habitat, prompting conspecifics to congregate in areas with abundant resources (Erbilgin, Powell and Raffa, 2003). This phenomenon can lead to the establishment of localised insect populations that exploit specific plant hosts.

Alarm pheromones represent a third category of pheromones, evoking responses that enhance individual or group survival (Hauser *et al.*, 2005). These chemical cues are often emitted when an organism is under threat, alerting nearby individuals to potential danger (Verheggen, Haubruge and Mescher, 2010). For instance, when honeybees are threatened, they release alarm pheromones that trigger an aggressive response from other members of the hive, ensuring collective defence (López-Incera *et al.*, 2021).

2.3 PRODUCTION AND DETECTION OF SEMIOCHEMICALS

Semiochemicals, the chemical messengers that orchestrate the intricate dialogues between organisms, are produced by a multitude of species across the biological spectrum (Bakthavatsalam, 2016). These compounds are emitted intentionally or unconsciously, shaping behaviours, interactions and ecological dynamics (El-Ghany, 2019). The production of semiochemicals is a sophisticated process deeply intertwined with an organism's physiology, genetics and environment. The production of semiochemicals is a multifaceted process that varies across different organisms, reflecting their evolutionary history, ecological niche and functional roles (Bakthavatsalam, 2016). Semiochemicals can originate from diverse sources within an organism's body, such as specialised glands, tissues or even metabolic pathways (Barbosa-Cornelio *et al.*, 2019). Many organisms possess specialised glands that synthesise and release semiochemicals (Tittiger and Blomquist, 2017). In insects, for instance, pheromones are often produced by glandular structures such as pheromone glands (Costa-Leonardo and Haifig, 2010). These glands are responsible for the synthesis and storage of pheromone compounds, which are then released in response to specific behavioural cues (Cardé and Millar, 2009). Similarly, plants possess glandular structures that release volatile organic compounds (VOCs) that can attract pollinators, repel herbivores or signal stress (Cardé and Millar, 2009). Additionally, semiochemicals can be by-products of metabolic processes within organisms. These compounds may serve secondary functions beyond their primary metabolic roles. For instance, the green leaf volatile (GLV) compounds released by plants are not produced for the purpose of communication, but their emission can alert neighbouring plants to the presence of herbivores, enabling the activation of defence mechanisms (Li, Zhu and Qin, 2012). Environmental factors, such as temperature, humidity and light, as well as genetic factors, play significant roles in influencing the production of semiochemicals (Figueiredo *et al.*, 2008; Hock *et al.*, 2014). This interplay of genetic and environmental factors contributes to the diversity of semiochemical profiles among populations (Sharma, Sandhi and Reddy, 2019).

The detection of semiochemical signals is equally intricate and essential for organisms to obtain valuable information from their surroundings. The mechanisms underlying the detection of semiochemicals are highly specialised, with receptors and sensory structures evolved to perceive specific compounds (Brezolin *et al.*, 2018). In insects, olfactory receptors play a central role in detecting semiochemicals. These receptors are located on sensory organs, such as antennae, and they are finely tuned to recognise specific compounds. Each receptor is specialised to detect a particular type of semiochemical, allowing insects to discriminate between different chemical cues (Leal, 2013). This specificity enables insects to accurately interpret the information carried by various semiochemicals. The activation of olfactory receptors triggers neuronal signals that are transmitted to the brain, resulting in the perception of an odour. The sensitivity of these receptors is remarkable, allowing insects to detect pheromones or other signals even when emitted in trace amounts (Fleischer and Krieger, 2018). Additionally, the interpretation of semiochemical signals often involves a contextual element. The same semiochemical might evoke different responses depending on the context in which it is perceived (Orlova and Amsalem, 2021). For instance, a pheromone emitted by a conspecific might signal mating readiness in one context but trigger an alarm response in another, depending on the physiological state or environmental conditions of the receiver (Tirindelli *et al.*, 2009; Renou, 2014). Furthermore, semiochemical detection often operates in conjunction with other sensory cues, such as visual or vibrational signals. This integration of multiple sensory modalities enhances an organism's ability to accurately perceive and respond to their environments. The production and detection of semiochemicals exemplify the remarkable adaptations that have arisen throughout evolutionary history, enabling organisms to communicate, interact and thrive in the intricate tapestry of ecological relationships.

2.4 ROLE OF SEMIOCHEMICALS IN COMMUNICATION

Semiochemicals, as chemical messengers, play a pivotal role in facilitating communication among organisms across diverse ecological contexts (Smart, Aradottir and Bruce, 2014). This communication extends beyond species boundaries, encompassing interactions between conspecifics and heterospecifics alike. Semiochemicals mediate an intricate web of relationships, guiding behaviours that range from attracting mates and deterring predators to locating food sources and coordinating social interactions (Bakthavatsalam, 2016). In the context of mate attraction, pheromones are exemplars of semiochemicals that underpin the dance of reproduction. These chemical signals carry species-specific information about an individual's readiness to mate, acting as invitations that traverse the air, guiding potential partners to one another (Li, 2016). Beyond reproduction, semiochemicals also serve as alarm signals, alerting neighbouring organisms to potential threats or danger. For instance, when a plant is attacked by herbivores, it can emit volatile compounds that act as a cry for help, signalling nearby plants to activate their defence mechanisms (Li, 2016). The intricate interplay between semiochemicals and communication highlights their role as information carriers, shaping behaviours that are essential for survival, reproduction and adaptation. This complex chemical language forms the foundation of ecological interactions, allowing organisms to navigate their environment and respond to dynamic challenges in a world rich with chemical cues.

2.5 APPLICATIONS OF SEMIOCHEMICALS IN PEST MANAGEMENT

The intricate world of semiochemicals has yielded innovative applications in the realm of pest management, offering sustainable and ecologically friendly solutions to mitigate the impact of pests on agricultural and natural ecosystems (Smart, Aradottir and Bruce, 2014). Harnessing the power of semiochemicals, pest management strategies have been refined to target specific pests with precision, minimising the use of synthetic pesticides and reducing environmental risks (Ali *et al.*, 2021). One notable application lies in the development of semiochemical-based traps and lures. Pheromones, a prominent class of semiochemicals, play a pivotal role in the mating behaviours of many insects. By synthetically producing and deploying these pheromones, researchers have created traps that attract and capture male insects, disrupting their reproductive patterns (Witzgall, Kirsch and Cork, 2010; Li, Zhu and Qin, 2012). This strategy not only reduces pest populations by curtailing mating success but also provides a means of monitoring pest abundance. These traps enable farmers to gauge the severity of infestations and make informed decisions about the timing and necessity of control measures. The precision of these traps reduces the reliance on broad-spectrum pesticides, minimising non-target impacts on beneficial insects and the environment (Cui and Zhu, 2016).

Another application of semiochemicals involves the concept of "push-pull" strategies (Khan *et al.*, 2012). This approach utilises both repellent and attractive semiochemicals to manipulate pest behaviours and movements within agricultural landscapes. Repellent compounds are strategically deployed to "push" pests away from valuable crops, creating a protective buffer. Concurrently, attractive semiochemicals are used to "pull" pests toward designated trap crops or areas where control measures are implemented (Cook, Khan and Pickett, 2007; Khan *et al.*, 2012). This innovative strategy not only minimises damage to main crops but also promotes the establishment of targeted control measures in specific zones, further reducing the environmental footprint of pest management efforts. Furthermore, the understanding of semiochemical communication has paved the way for the development of "attract-and-kill" techniques (Komala, Manda and Seram, 2021). In these systems, semiochemical lures are coupled with lethal agents, such as insecticides or pathogens. The lure attracts pests to a treated area where they come into contact with the lethal agent, resulting in mortality. This approach offers a localised and targeted means of pest control, minimising the dispersal of harmful substances in the environment while maximising their impact on the pest population (El-Ghany, 2019). In conclusion, the applications of semiochemicals in pest management underscore the potential of chemical ecology to revolutionise our approach to agricultural challenges. By exploiting the intricate language of chemical cues, these strategies offer effective alternatives to conventional pest control methods, aligning with ecological principles and promoting sustainable practices. As our understanding of semiochemical communication continues to deepen, the future holds the promise of even more innovative and refined approaches that harness the power of nature's chemical conversations to enhance pest management while safeguarding our ecosystems.

Application of semiochemicals extends beyond the realm of ecological interactions and into various industrial sectors in which their unique properties are harnessed for practical purposes. The industrial use of semiochemicals capitalises on their ability to influence behaviours and interactions, offering innovative solutions in fields ranging from agriculture to pest management (Heuskin *et al.*, 2011). In agriculture, semiochemicals have found utility in enhancing crop yields through techniques such as pheromone-based mating disruption. By releasing synthetic pheromones that saturate the environment, the mating patterns of pest insects can be disrupted, reducing their reproductive success and, subsequently, curbing their population growth (Bruce *et al.*, 2003). This approach provides an environmentally friendly alternative to conventional pesticide use, minimising ecological impacts and preserving beneficial insects. Additionally, semiochemicals play a pivotal role in precision agriculture, aiding in the monitoring and control of pest populations by targeting specific species while sparing non-target organisms (Pickett *et al.*, 2007).

Beyond agriculture, semiochemicals have been embraced in forestry for managing insect pests that threaten forest ecosystems. By deploying semiochemical-based traps and lures, foresters can monitor and mitigate the spread of destructive pests such as bark beetles (Gitau *et al.*, 2013). The industrial sector also benefits from semiochemicals' ability to attract or repel insects. In food production and storage, semiochemical-based traps help protect crops from post-harvest pests, ensuring the quality and safety of stored products (Li, Zhu and Qin, 2012). Additionally, semiochemicals are employed in integrated pest management programs for structural pest control, offering a sustainable approach that minimises the use of conventional pesticides in indoor environments (Mauchline, Hervé and Cook, 2018).

In the field of public health, semiochemicals play a crucial role in the control of disease vectors such as mosquitoes (Kline, 2007). Synthetic compounds that mimic the attractant cues of host animals or human hosts are used in mosquito traps, significantly reducing the prevalence of vector-borne diseases (Mafra-Neto and Dekker, 2019; Mweresa *et al.*, 2020). The utilisation of semiochemicals in this context not only safeguards public health but also mitigates the environmental and health risks associated with widespread pesticide use. As industries continue to explore and harness the potential of semiochemicals, their multifaceted applications hold the promise of innovative and sustainable solutions that bridge the gap between ecological understanding and industrial advancement.

2.6 CONCLUSION

In conclusion, the study of semiochemicals illuminates the complex language of chemical communication underlying ecological interactions among diverse organisms. Semiochemicals play multifaceted roles in mediating behaviours, guiding interactions and shaping ecological dynamics, highlighting their significance in nature. From plant emissions attracting pollinators to insect pheromones signalling mating readiness or alarm, semiochemicals are integral to ecological processes. Through exploring semiochemical production, detection and behavioural impacts, researchers uncover the complexity and implications of chemical communication. Moving forward, advancements in analytical techniques offer opportunities to identify and quantify elusive semiochemical compounds, enhancing our understanding of their roles. Integrating genomics and transcriptomics with semiochemical research can unveil the genetic basis of semiochemical production and reception, illuminating evolutionary aspects. As ecological contexts evolve due to environmental changes, understanding how semiochemical-mediated interactions adapt is essential. Innovations such as semiochemical-based pest management strategies benefit from comprehending mechanisms governing semiochemical production and detection. Exploring semiochemicals delves into the intricate world of chemical ecology, in which compounds transcend their nature to shape organisms' behaviours. As our understanding deepens, so does our appreciation for life's orchestration through chemical conversations. The future of semiochemical research promises to reveal new mysteries, uncover hidden connections and offer insights into nature's delicate balance of interactions.

REFERENCES

Abd El-Ghany, N. M. (2020) 'Pheromones and chemical communication in insects', *Pests, weeds and diseases in agricultural crop and animal husbandry production*. IntechOpen, pp. 16–30.

Ali, J. *et al.* (2021) 'Effects of cis-Jasmone treatment of brassicas on interactions with Myzus persicae Aphids and their parasitoid diaeretiella rapae', *Frontiers in Plant Science*, 12. doi: 10.3389/fpls.2021.711896.

Bakthatvatsalam, N., Subharan, K. and Mani, M. (2022) 'Semiochemicals and their potential use in pest management in horticultural crops', *Trends in horticultural entomology*, pp. 283–312.

Bakthavatsalam, N. (2016) 'Semiochemicals', *Ecofriendly pest management for food security*. Elsevier, pp. 563–611.

Baldwin, I. T. *et al.* (2006) 'Volatile signaling in plant-plant interactions: "Talking trees" in the genomics era', *Science*, 311(5762), pp. 812–815. doi: 10.1126/science.1118446.

Barbosa-Cornelio, R. *et al.* (2019) 'Tools in the investigation of volatile semiochemicals on insects: from sampling to statistical analysis', *Insects*, 10(8), p. 241.

Barbosa, P. and Letourneau, D. K. (1988) *Novel aspects of insect-plant interactions*. John Wiley & Sons.

Blassioli-Moraes, M. C. *et al.* (2019) 'Semiochemicals for integrated pest management', *Sustainable agrochemistry: a compendium of technologies*, pp. 85–112.

Brezolin, A. N. *et al.* (2018) 'Tools for detecting insect semiochemicals: a review', *Analytical and Bioanalytical Chemistry*, 410(17), pp. 4091–4108.

Brown Jr, W. L., Eisner, T. and Whittaker, R. H. (1970) 'Allomones and kairomones: transspecific chemical messengers', *Bioscience*, 20(1), p. 21.

Bruce, T. J. A. *et al.* (2003) 'cis-Jasmone treatment induces resistance in wheat plants against the grain aphid, sitobion avenae (Fabricius) (Homoptera: Aphididae)', *Pest Management Science*, 59(9), pp. 1031–1036. doi: 10.1002/ps.730.

Bruce, T. J. A. and Pickett, J. A. (2011) 'Perception of plant volatile blends by herbivorous insects–finding the right mix', *Phytochemistry*, 72(13), pp. 1605–1611.

Cardé, R. T. and Millar, J. G. (2009) 'Pheromones', *Encyclopedia of insects*. Elsevier, pp. 766–772.

Christensen, T. A. and Sorensen, P. W. (1996) 'Introduction: pheromones as tools for olfactory research', *Chemical Senses*, 21(2), pp. 241–243.

Cook, S. M., Khan, Z. R. and Pickett, J. A. (2007) 'The use of push-pull strategies in integrated pest management', *Annual Review of Entomology*, 52, pp. 375–400.

Costa-Leonardo, A. M. and Haifig, I. (2010) 'Pheromones and exocrine glands in Isoptera', *Vitamins & Hormones*, 83, pp. 521–549.

Cui, G. Z. and Zhu, J. J. (2016) 'Pheromone-based pest management in China: past, present, and future prospects', *Journal of Chemical Ecology*, 42, pp. 557–570.

Dicke, M. *et al.* (1990) 'Plant strategies of manipulating predatorprey interactions through allelochemicals: prospects for application in pest control', *Journal of Chemical Ecology*, 16, pp. 3091–3118.

Dicke, M. and Sabelis, M. W. (1988) 'Infochemical terminology: based on cost-benefit analysis rather than origin of compounds?', *Functional ecology*, pp. 131–139.

Divekar, P. A. *et al.* (2022) 'Plant secondary metabolites as defense tools against herbivores for sustainable crop protection', *International Journal of Molecular Sciences*, 23(5), p. 2690.

Dulac, C. and Torello, A. T. (2003) 'Molecular detection of pheromone signals in mammals: from genes to behaviour', *Nature Reviews Neuroscience*, 4(7), pp. 551–562.

El-Ghany, N. M. A. (2019) 'Semiochemicals for controlling insect pests', *Journal of Plant Protection Research*, 59(1).

Engelberth, J. *et al.* (2004) 'Airborne signals prime plants against insect herbivore attack', *Proceedings of the National Academy of Sciences of the United States of America*, 101(6), pp. 1781–1785. doi: 10.1073/pnas.0308037100.

Erbilgin, N., Powell, J. S. and Raffa, K. F. (2003) 'Effect of varying monoterpene concentrations on the response of Ips pini (Coleoptera: Scolytidae) to its aggregation pheromone: implications for pest management and ecology of bark beetles', *Agricultural and Forest Entomology*, 5(4), pp. 269–274.

Figueiredo, A. C. *et al.* (2008) 'Factors affecting secondary metabolite production in plants: volatile components and essential oils', *Flavour and Fragrance Journal*, 23(4), pp. 213–226.

Fleischer, J. and Krieger, J. (2018) 'Insect pheromone receptors–key elements in sensing intraspecific chemical signals', *Frontiers in Cellular Neuroscience*, 12, p. 425.

Gitau, C. W. *et al.* (2013) 'A review of semiochemicals associated with bark beetle (Coleoptera: Curculionidae: Scolytinae) pests of coniferous trees: a focus on beetle interactions with other pests and their associates', *Forest Ecology and Management*, 297, pp. 1–14. doi: 10.1016/j.foreco.2013.02.019.

Hansson, B. and Wicher, D. (2016) 'Chemical ecology in insects', *Chemosensory transduction*, pp. 29–45.

Hauser, R. *et al.* (2005) 'Alarm pheromones as an exponent of emotional state shortly before death–science fiction or a new challenge?', *Forensic Science international*, 155(2–3), pp. 226–231.

Heil, M. and Bueno, J. C. S. (2007) 'Within-plant signaling by volatiles leads to induction and priming of an indirect plant defense in nature', *Proceedings of the National Academy of Sciences of the United States of America*, 104(13), pp. 5467–5472. doi: 10.1073/pnas.0610266104.

Heuskin, S. *et al.* (2011) 'The use of semiochemical slow-release devices in integrated pest management strategies', *Base*, 15.

Hickman, D. T. *et al.* (2021) 'Allelochemicals as multi-kingdom plant defence compounds: towards an integrated approach', *Pest Management Science*, 77(3), pp. 1121–1131.

Hock, V. *et al.* (2014) 'Establishing abiotic and biotic factors necessary for reliable male pheromone production and attraction to pheromones by female plum curculios Conotrachelus nenuphar (Coleoptera: Curculionidae)', *The Canadian Entomologist*, 146(5), pp. 528–547.

Jacobson, M. (2012) *Insect sex pheromones*. Elsevier.

Jilani, G. *et al.* (2008) 'Allelochemicals: sources, toxicity and microbial transformation in soil—a review', *Annals of Microbiology*, 58, pp. 351–357.

Karban, R. (2015) *Plant sensing and communication*. University of Chicago Press.

Kessler, A. and Baldwin, I. T. (2001) 'Defensive function of herbivore-induced plant volatile emissions in nature', *Science*, 291(5511), pp. 2141–2144. doi: 10.1126/science.291.5511.2141.

Khan, Z. *et al.* (2012) 'Push–pull technology: a conservation agriculture approach for integrated management of insect pests, weeds and soil health in Africa: UK Government's foresight food and farming futures project', *Sustainable intensification*. Routledge, pp. 162–170.

Kline, D. L. (2007) 'Semiochemicals, traps/targets and mass trapping technology for mosquito management', *Journal of the American Mosquito Control Association*, 23(sp2), pp. 241–251.

Komala, G., Manda, R. R. and Seram, D. (2021) 'Role of semiochemicals in integrated pest management', *International Journal of Entomology Research*, 6(2), pp. 247–253.

Leal, W. S. (2013) 'Odorant reception in insects: roles of receptors, binding proteins, and degrading enzymes', *Annual Review of Entomology*, 58, pp. 373–391.

LeBlanc, H. N. (2021) *Olfactory stimuli associated with the different stages of vertebrate decomposition and their role in the attraction of the blowfly Calliphora vomitoria (Diptera: Calliphoridae) to carcasses*. University of Derby (United Kingdom).

Li, P., Zhu, J. and Qin, Y. (2012) 'Enhanced attraction of Plutella xylostella (Lepidoptera: Plutellidae) to pheromone-baited traps with the addition of green leaf volatiles', *Journal of Economic Entomology*, 105(4), pp. 1149–1156.

Li, T. (2016) 'Neighbour recognition through volatile-mediated interactions', *Deciphering chemical language of plant communication*. Springer, pp. 153–174.

López-Incera, A. *et al.* (2021) 'Honeybee communication during collective defence is shaped by predation', *BMC Biology*, 19(1), pp. 1–16.

Macias, F. A. *et al.* (2007) 'Allelopathy—a natural alternative for weed control', *Pest Management Science: Formerly Pesticide Science*, 63(4), pp. 327–348.

Mafra-Neto, A. and Dekker, T. (2019) 'Novel odor-based strategies for integrated management of vectors of disease', *Current Opinion in Insect Science*, 34, pp. 105–111.

Mauchline, A. L., Hervé, M. R. and Cook, S. M. (2018) 'Semiochemical-based alternatives to synthetic toxicant insecticides for pollen beetle management', *Arthropod-Plant Interactions*, 12(6), pp. 835–847.

Meiners, T. and Hilker, M. (2000) 'Induction of plant synomones by oviposition of a phytophagous insect', *Journal of Chemical Ecology*, 26(1), pp. 221–232.

Mithöfer, A., Boland, W. and Maffei, M. E. (2009) 'Chemical ecology of plant-insect interactions', *Molecular aspects of plant disease resistance*. Wiley-Blackwell, pp. 261–291.

De Moraes, C. M. *et al.* (1998) 'Herbivore-infested plants selectively attract parasitoids', *Nature*, 393(6685), pp. 570–573.

Mweresa, C. K. *et al.* (2020) 'Use of semiochemicals for surveillance and control of hematophagous insects', *Chemoecology*, 30, pp. 277–286.

Nault, L. R., Edwards, L. J. and Styer, W. E. (1973) 'Aphid alarm pheromones: secretion and reception', *Environmental Entomology*, 2(1), pp. 101–105.

Orlova, M. and Amsalem, E. (2021) 'Bumble bee queen pheromones are context-dependent', *Scientific Reports*, 11(1), p. 16931.

Paré, P. W. and Tumlinson, J. H. (1999) 'Plant volatiles as a defense against insect herbivores', *Plant Physiology*, 121(2), pp. 325–331. doi: 10.1104/pp.121.2.325.

Pickett, J. A. *et al.* (2007) 'Plant volatiles yielding new ways to exploit plant defence', *Chemical ecology*, pp. 161–173. doi: 10.1007/978-1-4020-5369-6_11.

Qi, S.-S. *et al.* (2020) 'Allelopathy confers an invasive Wedelia higher resistance to generalist herbivore and pathogen enemies over its native congener', *Oecologia*, 192, pp. 415–423.

Regnault-Roger, C. and Philogène, B. J. R. (2008) 'Past and current prospects for the use of botanicals and plant allelochemicals in integrated pest management', *Pharmaceutical Biology*, 46(1–2), pp. 41–52.

Renou, M. (2014) 'Pheromones and general odor perception in insects', *Neurobiology of Chemical Communication*, 1, pp. 23–56.

Salerno, G. *et al.* (2013) 'Short-range cues mediate parasitoid searching behavior on maize: the role of oviposition-induced plant synomones', *Biological Control*, 64(3), pp. 247–254.

Sbarbati, A. and Osculati, F. (2006) 'Allelochemical communication in vertebrates: kairomones, allomones and synomones', *Cells Tissues Organs*, 183(4), pp. 206–219.

Schnee, C. *et al.* (2006) 'The products of a single maize sesquiterpene synthase form a volatile defense signal that attracts natural enemies of maize herbivores', *Proceedings of the National Academy of Sciences of the United States of America*, 103(4), pp. 1129–1134. doi: 10.1073/pnas.0508027103.

Scolari, F. *et al.* (2021) 'Tephritid fruit fly semiochemicals: current knowledge and future perspectives', *Insects*, 12(5), p. 408.

Sharma, A., Sandhi, R. K. and Reddy, G. V. P. (2019) 'A review of interactions between insect biological control agents and semiochemicals', *Insects*, 10(12), p. 439.

Shorey, H. H. (1973) 'Behavioral responses to insect pheromones', *Annual Review of Entomology*, 18(1), pp. 349–380.

Shorey, H. H. (2013) *Animal communication by pheromones*. Academic Press.

Smart, L. E., Aradottir, G. I. and Bruce, T. J. A. (2014) 'Role of semiochemicals in integrated pest management', *Integrated pest management*. Elsevier, pp. 93–109.

Tirindelli, R. *et al.* (2009) 'From pheromones to behavior', *Physiological Reviews*, 89(3), pp. 921–956.

Tittiger, C. and Blomquist, G. J. (2017) 'Pheromone biosynthesis in bark beetles', *Current Opinion in Insect Science*, 24, pp. 68–74.

Tlak Gajger, I. and Dar, S. A. (2021) 'Plant allelochemicals as sources of insecticides', *Insects*, 12(3), p. 189.

Verheggen, F. J., Haubruge, E. and Mescher, M. C. (2010) 'Alarm pheromones—chemical signaling in response to danger', *Vitamins & Hormones*, 83, pp. 215–239.

Witzgall, P., Kirsch, P. and Cork, A. (2010) 'Sex pheromones and their impact on pest management', *Journal of Chemical Ecology*, 36(1), pp. 80–100.

Yew, J. Y. and Chung, H. (2015) 'Insect pheromones: an overview of function, form, and discovery', *Progress in Lipid Research*, 59, pp. 88–105.

CHAPTER 3

Chemoreception in Insects

3.1 INTRODUCTION

Chemical communication serves as a fundamental cornerstone of inter-actions within the vast tapestry of the natural world (Baeckens, 2019). Across diverse ecological niches, organisms have harnessed the power of chemical cues to navigate their environment, defend against predators, locate mates and secure resources crucial for survival (Ferrari, Wisenden and Chivers, 2010). This complex interplay of volatile compounds, phero-mones and chemical signals orchestrates a dynamic symphony that shapes the dynamics of ecosystems and steers the behaviour and persistence of species (Schuman, 2023; Zu et al., 2023). Chemical ecology, a multidisci-plinary domain, seeks to unravel these intricate chemical dialogues and, in doing so, uncover their profound implications for the natural world (Cardé and Bell, 1995; Ahmad, Aslam and Razaq, 2004). Within this complex field of chemical communication, insects emerge as virtuosos, exemplifying the diverse and remarkable strategies that have evolved to navigate their intricate world (Wilson, 1965; Hansson and Stensmyr, 2011). As the most abundant and diverse group of animals, insects have evolved an aston-ishing array of chemoreceptive mechanisms that enable them to perceive and decode chemical cues in their surroundings (Murdoch, Evans and Peterson, 1972; Hansson and Stensmyr, 2011; Hansson and Wicher, 2016). These cues play pivotal roles, from guiding mating rituals to enabling the

DOI: 10.1201/9781003479857-3

location of suitable host plants and prey (Ryan, 2002a). Indeed, insects' reliance on their chemosensory abilities is woven intricately into their very essence, shaping their behaviours, reproductive strategies and ecological roles (Ryan, 2002a, 2002c; Hansson and Wicher, 2016).

This chapter explores the intriguing world of insect chemoreception, focusing on the mechanisms that allow insects to understand their chemical environment. We investigate the anatomical adaptations that help insects detect these chemical cues, including specialised structures such as sensilla, antennae and mouthparts. Additionally, we examine the molecular mechanisms involved in chemoreception, such as receptor proteins and signal transduction pathways, which enable insects to interpret their chemical surroundings.

3.2 CHEMORECEPTION MECHANISMS IN INSECTS

Chemoreception, the process by which organisms detect and interpret chemical cues from their environment, plays a pivotal role in the survival and success of insects (Dethier, 1957; Mathis and Crane, 2017). This intricate sensory mechanism enables insects to perceive and respond to a myriad of chemical stimuli, guiding their behaviours such as foraging, mating and avoiding danger (Depetris-Chauvin, Galagovsky and Grosjean, 2015; Hansson and Wicher, 2016; Mathis and Crane, 2017; Zhengyan Wang et al., 2023). This section explores the fundamental chemoreceptive processes of olfaction and gustation as well as the specialised anatomical structures that facilitate chemical detection in insects.

3.2.1 Olfaction

Olfaction, the detection of volatile chemicals in the air, is a cornerstone of insect chemoreception. Insects possess an impressive diversity of olfactory receptors that allow them to detect a wide range of chemical compounds (Missbach et al., 2014; Wicher and Miazzi, 2021). Olfactory receptors are typically located in specialised sensory structures called sensilla, which are distributed across various parts of the insect body, most notably on the antennae and mouthparts (Wicher and Miazzi, 2021). These sensilla serve as specialised sites for chemical reception, housing the receptor proteins that interact with odorant molecules (Zacharuk, 1980; Steinbrecht, 2007).

Antennae, in particular, are vital sensory organs for olfaction (Sachse and Krieger, 2011; Böröczky et al., 2013). Insects have evolved antennal structures that maximise surface area for chemical sampling. The antennae of some species are adorned with various types of sensilla, each tuned to detect specific chemical cues (Schneider, 1964; Schneider, Price and Moore, 1998; Hansson and Stensmyr, 2011). For instance, grooved peg sensilla are found on the antennae of moths and butterflies, enabling them to detect pheromones released by conspecifics for mating purposes (Rani et al., 2021). Other sensilla, such as chaetic sensilla, respond to a wide array of odorants, allowing insects to perceive environmental cues such as food sources, predators or oviposition sites (Nakanishi et al., 2009; Missbach et al., 2020).

3.2.2 Gustation

Gustation, or chemoreception through contact with chemical stimuli, is equally essential for insects' interactions with their surroundings (Chapman, 2003; Bell and Cardé, 2013). Mouthparts, including labial palps and tarsi, often bear chemosensory sensilla that facilitate the perception of chemicals through physical contact (de Brito Sanchez, 2011; Param et al., 2022). These sensilla are strategically positioned to come into direct contact with food sources, plant surfaces or other substances of interest (Lee and Strausfeld, 1990; Barrozo, 2019). By tapping into gustatory cues, insects can assess the suitability of potential food items, identify toxic compounds and navigate complex chemical landscapes (Patt et al.. 2014; Shields and Shields, 2021).

3.2.3 Specialised Structures for Chemical Detection

Insects have evolved a remarkable array of specialised structures to enhance their chemoreceptive abilities. Sensilla are at the forefront of this adaptation, existing in a multitude of shapes and sizes, each adapted to detect specific types of chemical cues (Zacharuk, 1980). Sensilla are often equipped with sensory neurons that transduce chemical signals into electrical impulses, which are then transmitted to the insect's nervous system for processing (Steinbrecht, 2007). In addition to antennae and mouthparts, some insects possess other chemoreceptive structures. For instance, mosquitoes have specialised sensilla on their proboscis that aid in locating blood vessels, while bees have sensilla on their antennae that play a role in detecting floral scents (Amer and Mehlhorn, 2006; Fialho et al., 2014; Ren et al., 2023). In certain cases, insects may also have chemoreceptors on their legs or even abdominal segments, expanding their chemical sensory range. In conclusion, the chemoreception mechanisms in insects encompass olfaction and gustation, enabled by a rich diversity of sensory structures. These structures, including sensilla on antennae, mouthparts and other body parts, enable insects to detect a wide spectrum of chemical cues, shaping their interactions with the environment and guiding behaviours critical for survival and reproduction. The next section will delve deeper into the molecular mechanisms underlying these processes, shedding light on the intricate world of insect chemoreception.

3.3 CHEMICAL SENSES IN INSECTS

It is widely recognised that insects' chemical senses can be broadly classified into three categories, corresponding to smell, taste and the common chemical sense observed in vertebrates (Tucker and Smith, 1969). However, the criteria for this classification are as delicate as those applied to vertebrates, prompting a re-examination of traditional points used to differentiate gustatory and olfactory senses in light of recent insights into invertebrate chemoreceptors. Challenges arise in using anatomical and topographical factors as valid criteria in insects. The distribution of chemoreceptors, a topic debated since 1798, faces scrutiny because of ongoing discoveries of new loci and questionable experimental techniques. Assigning specific areas of the body to each chemical sense is currently inadvisable, especially as end organs for both senses often coexist in the same segments of an appendage (Frings and Frings, 1949). For example, in honeybees, both antennal and labial nerves converge at the deutocerebrum, challenging traditional distinctions (Anton and Homberg, 1999). In contrast to vertebrates, in which olfactory and taste receptors have distinct nerves, insects exhibit convergence (Hansson and Stensmyr, 2011). Fibres from both olfactory and gustatory receptors enter the same nerve and reach the same brain lobe (Eisthen, 2002). The distinction between taste and smell becomes evident when examining concentration limits. For instance, quinine hydrochloride, a potent taste substance for humans, is perceived at concentrations around 1.5×10^{-7} M, while aquatic beetles respond at dilutions of 1.25×10^{-6} M. In contrast, an effective olfactory stimulus for humans, a mercaptan, is perceived at approximately 9×10^{-13} M, with comparable values lacking for insects. Sensitivity experiments reveal that tarsal receptors in butterflies and flies respond to sucrose concentrations as low as 9.8×10^{-4} M and 3.9×10^{-5} M, respectively (Dethier and Chadwick, 1948). Challenges in drawing strict distinctions persist, but insects exhibit dual receptors – olfactory for volatile compounds and gustatory for liquids or solutions, mirroring vertebrates.

Insects unequivocally possess a distinct chemoreceptor category, the common chemical sense, responding to high concentrations of irritating compounds such as ammonia, chlorine and essential oils (Minnich, 1929). Failure to acknowledge this sense led to contradictory findings in experiments locating olfactory organs. Certain compounds stimulate both olfactory and common chemical senses, as seen in insects conditioned to weak odours of essential oils (Schoonhoven, 2018). Quantitative studies are limited, but observations suggest a heightened response threshold, typically manifesting as an avoidance reaction. The specific localisation of common chemical sense receptors remains undiscovered. Contact chemoreception, considered a common chemical sense, lacks wholly satisfactory criteria for separation from other senses in certain aquatic and subterranean insects (Dethier and Chadwick, 1948; Chapman, 2003). The function primarily revolves around feeding, oviposition and influencing resting places. Loci of receptors include antennae, mouthparts, legs and ovipositors in various species (Frings and Frings, 1949). However, uncertainties persist regarding the nature of receptors in insects. Attempts to match structure and specific function have been unsuccessful, with appendages often serving multiple functions. While some studies suggest success in stimulating specific hairs, experimental verification of presumptive end organs remains challenging. Efforts to assign function based solely on morphology, without experimental support, obscure the situation. Notably, the presence of "olfactory pores" (sensilla campaniformia) on tarsal segments is likely unrelated to chemoreception, recognised as proprioceptive organs (Dethier and Chadwick, 1948; Hu et al., 2018; Acevedo Ramos et al., 2020). Ongoing challenges underscore the need for further experimental inquiry into the specific nature of receptors involved in contact chemoreception.

3.3.1 Role of Cuticle in Chemoreception

Insects, boasting a chitin-protein complex known as the cuticle, play a crucial role in chemoreception such as taste and smell perception. This protective layer is coated with wax to prevent desiccation. Sensory cells responsible for taste and smell are housed within cuticular structures such as hairs, pegs and flat surfaces, particularly on antennae, mouthparts, legs and the ovipositor (Ozaki and Wada-Katsumata, 2010). These structures, collectively termed sensilla, possess modified cuticular regions housing pores facilitating the entry of chemicals. To maintain water conservation and sensory cell functionality, these pores ensure a water–protein pathway to the cell membrane (Ali et al., 2015). Reconstruction of a typical mouthpart gustatory sensillum in a caterpillar reveals intricate cellular details, showcasing the importance of such structures in the food selection process. Chemoreceptive cells in sensilla are modified cilia, a common feature across various sensory cells in animals. While both taste and olfactory sensilla share structural features, differences lie in how chemicals enter the system (Wang and Dai, 2017). Gustatory sensilla have a single protective pore at the tip, limiting desiccation and potentially restricting chemical types. In contrast, olfactory sensilla feature multiple pores, repurposing pore canals in the cuticle for stimuli access to sensory dendrites (King and Gunathunga, 2023). The intricacies of the insect cuticle play a pivotal role in mediating the interaction between the external environment and the sensory cells responsible for taste and smell perception.

3.4 CHEMOSENSORY MECHANISMS IN INSECTS

In exploring the intricate world of chemosensory perception in insects, the odour path serves as a captivating starting point. The journey of odour molecules from encountering the waxy cuticular surface to navigating pore canals and finally reaching the dendritic surface of sensory cells unveils a nuanced process (Mitchell, 2009). Recent revelations involving proteins such as pheromone binding proteins and general odorant binding proteins contribute to the understanding of shuttling odour molecules through extracellular spaces.

- *Chemical-to-Electrical Transduction:* Drawing parallels with vertebrates, the process of chemical-to-electrical transduction in insects involves specific receptor molecules, amplification steps, ion channels and deactivation systems. Recent advances, including patch-clamp studies and genetic analyses, shed light on the components of this transduction system, showcasing both similarities and differences with vertebrate arrangements.

- *Chemosensory Coding at the Periphery:* The periphery of insect chemosensory coding introduces the dual challenge of filtering out non-critical stimuli while maintaining sensitivity to biologically relevant ones. From simpler labelled-line codes observed in pheromone systems to more complex across-fibre patterns seen in plant-feeding insects, the encoding of stimuli reflects the intricate strategies employed by insects to make sense of their chemical environment.

- *Central Processing of Chemosensory Input:* Over the past two decades, studies have delved into the central processing of insect olfactory systems, particularly in the context of pheromones. The organisational distinctions between olfactory and gustatory systems emerge, with olfactory systems exhibiting a glomerular pattern and gustatory systems lacking such a structured arrangement. These structural differences hint at potential divergences in coding and evolutionary trajectories across the insect kingdom.

3.5 MOLECULAR BASIS OF CHEMORECEPTION IN INSECTS

Like other living organisms, insects gather information about their surroundings through various senses, including vision, hearing, smell and taste. These sensory systems rely on specialised tissue structures called receptors, which are sensitive to specific physical stimuli (Dethier and Chadwick, 1948; Mitchell, 2009). Receptor cells, such as neurons, undergo signal transduction to convert stimuli into nerve impulses, utilising receptor molecules that either activate ion channels (ionotropic) or initiate membrane signalling cascades through specialised G proteins (metabotropic). Multicellular organisms combine these mechanisms for sophisticated environmental perception.

- *Common Principles of Chemoreception*: Chemoreception is vital for insects to perceive and understand their environment. It involves recognising chemical stimuli such as tastes, smells and various substances, contributing to aspects such as food quality, predator detection and social interactions (Ryan, 2002a). Insects employ diverse chemosensory systems, with chemosensory transduction in neurons converting biochemical signals into electrical signals.

- *Chemoreceptors in Insects*: Insects rely on olfactory sensory neurons (OSNs) for odour perception, located in appendages such as the forehead, antennae and maxillary palps (Boeckh, Kaissling and Schneider, 1965). Three types of receptors facilitate insect smell: odorant receptors (ORs), ionotropic receptors (IRs) and specialised gustatory receptors (GRs) (Missbach et al., 2014; Wicher and Miazzi, 2021). ORs, as heterodimers, recognise food odours and pheromones. IRs, homologous to ionotropic glutamate receptors, are sensitive to acids, amines and aldehydes. GRs, specific to carbon dioxide, form heterodimers activating phospholipase C and initiating Transient Receptor Potential (TRP)-family ion channels (Wicher and Miazzi, 2021).

- *Gustatory Receptors in Insects – Unravelling Taste Sensitivity*: Insects have a sophisticated taste sensory system primarily located on their legs and wings through structures called taste sensilla (Hallem, Dahanukar and Carlson, 2006; Freeman, Wisotsky and Dahanukar, 2014). These sensilla house gustatory receptor neurons (GRNs) with GRs in their dendrites. The GR gene family discovery marked a breakthrough, including 68 GR genes in the *Drosophila* genome. These genes categorise into groups responding to bitter and salty tastes or exclusively to sweet tastes. Bitter taste receptors, crucial for threat detection, exhibit sensitivity to about 35 GR genes. Core subunits of multimeric bitter taste GR receptors are believed to be GR33a and GR66a (Hallem, Dahanukar and Carlson, 2006).

- *Gustatory Signal Transduction:* The signalling pathways of GRs in insects remain incompletely understood, primarily due to challenges in studying gustatory neurons using electrophysiological methods and difficulties in expressing GRs in heterologous systems (Xu, Zhang and Anderson, 2012). The GR43a-like clade, a family of receptors involved in fructose taste perception, offers insights into ionotropic homosubunit chemoreceptors. Successful heterologous expression of genes such as BmGR9 from silkworm in human embryonic kidney cells has shed light on their ligand-gated cation channel properties (Xu, Zhang and Anderson, 2012; Xu, 2020). Research into insect gustatory receptors is continuously evolving, highlighting the diverse repertoire of receptors in different insect species. Members of the GR43a-clade offer ample opportunities for detailed studies across various model systems. Recent findings on receptors such as TcGR20 in *Tribolium castaneum*, sensitive to sorbitol and mannitol, open doors for potential advancements in electrophysiological instruments based on these receptors (Sokolinskaya et al., 2020).

Insect and vertebrate chemoreceptors differ in gene expression strategies (Kaupp, 2010). In *Drosophila*, olfactory neurons express two olfactory receptor genes each, lacking the feedback principle suppressing nonfunctional receptors. Eusocial insects exhibit differential chemoreceptor gene expression based on sex, age and caste, influencing social behaviours such as polyethism and community building (Zhang et al., 2016; Treanore, Derstine and Amsalem, 2021). Insects possess a complex chemoreception system with proteins from three superfamilies: ORs, IRs and GRs. The system's architecture reflects evolutionary adaptations, allowing insects to precisely respond to external chemical cues. Complementary sensitivity and unequal sensitisation abilities of GRs and IRs contribute to this accuracy. While an ionotropic signal transduction pathway is common, substantial gaps persist, and ongoing research aims to provide a comprehensive "chemoreceptor map," unravelling the intricacies of insect chemosensory systems.

3.6 NUTRIENT SENSATION AND PERCEPTION IN INSECTS

Understanding the intricate interplay between taste, metabolism and behaviour is pivotal for insects, influencing crucial aspects of their survival and reproduction. Navigating the complexities of nutritional assessment in diverse food sources, insects rely on taste to decode information about non-volatile nutrients (Rogers and Newland, 2003; Borkakati et al., 2019). While olfactory cues aid in locating food, taste, conveyed through a diverse array of substances, is essential for unravelling chemical intricacies. Evolutionary adaptations have equipped insects with receptive mechanisms that respond to internal and external cues, guiding decisions on food consumption. Hunger and neuromodulation regulate selectivity, with stretch receptors in the foregut acting as initial controllers of hunger. Post-consumption, internal nutrient receptors offer insights into nutrient profiles (Rogers and Newland, 2003).

Examining the diverse world of chemoreceptors in insects, including olfactory receptors, GRs and IRs, unveils their critical role in nutrient perception (Dethier and Chadwick, 1948; Mathis and Crane, 2017). While specific ligands are identified primarily for sugar receptors (SRs), the species-specific diversity of chemoreceptor repertoires is influenced by ecological and evolutionary factors. Evolutionary pathways underscore rapid diversification, particularly in olfactory receptors and GRs, shaping adaptations to nutritional niches (Auer et al., 2022). Bitter receptor diversity, exemplified in species such as *Bombyx mori* and *Helicoverpa armigera*, correlates with the evaluation of host plant defence (Ryan, 2002a, 2002b). Similarly, the number of SRs in insects such as *Drosophila* and bees aligns with dietary sugar diversity, reflecting their ecological roles in food selection and reproduction (Simcock, 2015; Melvin et al., 2018).

3.6.1 Diversity in Organs and Sensilla Distribution

Insects possess a gustatory system more diverse than that of mammals, utilising various contact chemoreceptive organs such as antennae, mouthparts and tarsi (Amrein and Thorne, 2005; Schoonhoven, 2018). Gustatory sensilla, varying in size and shape, contain GRNs interacting with chemosensory proteins. Sensilla, equipped with mechanosensory neurons, contribute to the assessment of food texture. Gustatory sensitivity relies on sensilla number and GRN receptor types, exhibiting variations within species and across body parts. The translation of gustatory information into action involves sweet gustatory projection neurons originating in the suboesophageal zone (SEZ), influencing proboscis extension responses (PER) (Li and Montell, 2021; King and Gunathunga, 2023). Despite advancements, higher order processing of contact chemoreceptive signals remains a realm with knowledge gaps.

3.6.2 Chemosensory Perception of Nutrient

Insects, requiring diverse macro- and micronutrients, showcase a sophisticated chemosensory system crucial for nutrient perception (Delompré et al., 2019; Ruedenauer et al., 2023). The reception of proteins/amino acids, sugars, fats/fatty acids, micronutrients and bitter tastes involves intricate adaptations. Protein/amino acid reception, vital for growth and functions, exhibits species-specific diversity in amino acid sensing influenced by foraging ecology. Sugar perception, pivotal for energy regulation, engages various receptors including SRs and IRs. Fat/fatty acid sensing, essential for energy storage, varies across species, influencing feeding choices (Delompré et al., 2019). Micronutrient reception, a less-explored domain, involves concentration-dependent activation of receptors. Bitter tastes, signalling toxicity, prompt avoidance, with bitter substances inhibiting sweet receptors. In essence, insect chemosensory systems reflect nuanced adaptations for precise nutrient perception and ecological niche specialisation (Delompré et al., 2019; Crumière et al., 2020).

The intricate mechanisms governing insect nutrient sensation underscore a tapestry of knowledge gaps and adaptations. Concentration-dependent activation and inhibition of receptors contribute to the complexity, suggesting the existence of a nuanced perception hierarchy (Schoonhoven, 1987; Ruedenauer et al., 2023). This adaptive hierarchy aligns with the insect's physiological state, prioritising essential nutrients (Douglas, 2003; Behmer and Joern, 2012). Insects dynamically adjust their receptive repertoire, enhancing sensitivity or signal amplification based on immediate needs (Wicher and Miazzi, 2021). The accuracy of nutrient perception exhibits variability among substances, contexts and species, highlighting a hierarchical focus on key beneficial or detrimental compounds (Wendin and Nyberg, 2021; Ruedenauer et al., 2023). This strategic prioritisation is paramount, especially when navigating complex food mixtures, enabling insects to efficiently assess food quality and optimise their foraging strategies (Schoonhoven, 1987; Rogers and Newland, 2003; Joseph and Carlson, 2015).

3.7 MICROBIAL INFLUENCE ON CHEMORECEPTION IN INSECTS

Chemoreception in insects is significantly influenced by microbial symbionts, as highlighted in recent studies (Engl and Kaltenpoth, 2018; Zhengyan Wang et al., 2023). The impact extends to various aspects of insect behaviour, from attracting natural enemies to influencing mating preferences. This modulation of host chemoperception by symbionts, including pathogens, has been observed to alter sensitivity to semiochemicals and influence behaviours such as oviposition preferences (Chakraborty and Roy, 2021). Throughout their evolutionary journey, insects have developed intricate associations with microbial symbionts, resulting in a mutualistic relationship shaped by coevolution. This symbiotic interaction profoundly influences insect chemoreception and provides insights into the coevolutionary dynamics of insect–microbe associations (Chakraborty and Roy, 2021). Understanding this relationship holds potential for advancing insect control strategies and species conservation efforts (Zhengyan Wang et al., 2023).

The association between microbial symbionts and insect chemoreception is primarily observed in insects harbouring endosymbionts or pathogens. Some symbionts negatively impact insect olfactory sensitivity, influencing mate acceptance and host responses to environmental cues (Mondal et al., 2023; Zhengyan Wang et al., 2023). For example, the transfer of the endosymbiont *Wolbachia pipientis* in parasitoid wasps and the infection of honeybees with deformed wing virus (DWV) result in reduced olfactory sensitivity, potentially due to energy trade-offs between immune reactions and chemoreception. Conversely, certain symbionts enhance the sensitivity of insect chemoreception (Zhengyan Wang et al., 2023). Studies with fruit flies (*Drosophila melanogaster*) and the whitefly *Bemisia tabaci* indicate a positive correlation between microbial load and chemotaxis response toward odorants. The endosymbiotic bacterium *Wolbachia*, for instance, increases the olfactory responsiveness of infected *Drosophila simulans*. This duality in symbiotic impact suggests a dynamic interplay between microbial symbionts and host chemoreception, influencing insect behaviour based on specific ecological contexts (Santos-Garcia et al., 2020; Giorgini et al., 2023; Zhengyan Wang et al., 2023).

The manipulation of insect chemosensory behaviour by microbial symbionts holds evolutionary significance, fostering mutual benefits for both partners (Mondal et al., 2023; Zheng-yan Wang et al., 2023). Leafcutter ants, for instance, exhibit plant preferences driven by the need to protect symbiotic fungi from harmful compounds, highlighting how symbiont-mediated chemoreception contributes to coevolutionary dynamics. Infections with parasites or pathogens trigger changes in insect feeding habits and chemosensory behaviour, influencing population dynamics. The alteration of host chemosensitivity induced by infections impacts aggregation and dispersal behaviours, contributing to population segregation (Heine et al., 2018; Bruner-Montero et al., 2021). This phenomenon is exemplified by locusts infected with the microsporidian parasite *Paranosema locustae*, in which suppressed gregariousness aids the survival and development of locust populations. Similarly, the infection of pea aphids with the fungal pathogen *Erynia neoaphidis* alters their sensitivity to alarm pheromones, impacting intra-population transmission and dispersal to healthy populations (Roy, Pell and Alderson, 1999; Feng et al., 2015; Moyano, Croce and Scolari, 2023). In summary, the intricate interplay between microbial symbionts and insect chemoreception shapes individual behaviours and contributes to broader ecological and evolutionary outcomes (Noman et al., 2020). Understanding this inter-dependence offers avenues for further research on coevolutionary dynamics and provides valuable insights for developing targeted strategies in insect control and conservation.

3.8 ANATOMICAL ADAPTATIONS FOR CHEMICAL SENSING

3.8.1 Sensory Structures and Their Distribution on the Insect Body

Insects, with their remarkable diversity and ecological success, have evolved an extraordinary array of sensory structures that enable them to perceive and interpret chemical cues from their environment (Schroeder et al., 2018; Wikantyoso et al., 2022). These sensory structures, known as sensilla, are specialised cuticular structures distributed across the insect body. Sensilla are sensory organs that house chemoreceptor cells responsible for detecting specific volatile chemicals or contact chemicals through gustation (Steinbrecht, 2007; Lucas, Montagné and Jacquin-Joly, 2022). The distribution of these sensilla varies among different insect species, reflecting their ecological needs and selective pressures (Cassau et al., 2022; Wang et al., 2022). The types and distribution of sensilla often align with the specific functions they serve. For instance, trichoid sensilla, hair-like structures found on various body parts, are commonly associated with olfactory functions. Mechanoreceptive sensilla, such as campaniform sensilla, are found on appendages and body segments and may play a role in integrating sensory inputs. These sensory structures form intricate networks that collectively contribute to an insect's chemical sensory capabilities (Cassau et al., 2022; Wang et al., 2022).

3.8.2 Morphological Variations in Antennae and Mouthparts for Specific Functions

Among the various anatomical adaptations for chemical sensing, antennae and mouthparts stand out as critical components. Antennae, often referred to as the insect's "olfactory organs," display remarkable morphological diversity across species (Callahan, 1975; Faucheux, Kristensen and Yen, 2006). Elongated, segmented and adorned with sensilla of various types, antennae serve as primary chemoreceptive structures. Different types of sensilla on antennae, such as trichoid, basiconic and coeloconic sensilla, are associated with different types of chemical cues, enabling insects to detect a wide range of volatile compounds (Gainett et al., 2017; Song et al., 2017).

Mouthparts, while primarily associated with feeding and manipulating food, also contribute to chemoreception (Wikantyoso et al., 2022). Morphological variations in mouthparts are linked to specific feeding habits and ecologies. For instance, piercing-sucking mouthparts of blood-feeding insects such as mosquitoes possess specialised sensilla that aid in locating hosts based on chemical cues. Similarly, the mouthparts of herbivorous insects, such as caterpillars, are equipped with sensilla to detect plant-specific chemicals that guide feeding and oviposition behaviours (Wang, Li and Dai, 2019; Zahran, Sawires and Hamza, 2022).

3.8.3 Integration of Chemosensory Information with Other Sensory Modalities

Insects rely on a multimodal sensory system, in which chemosensory information is often integrated with other sensory inputs to form a comprehensive perception of their environment (Thiagarajan and Sachse, 2022). The integration of visual, tactile and auditory cues with chemical cues enhances an insect's ability to make accurate and adaptive decisions. For example, some insects use visual cues to locate floral resources, and their chemosensory system detects floral volatiles to confirm the presence of nectar or pollen (Wessnitzer and Webb, 2006). The integration of sensory modalities is particularly evident in social insects such as bees and ants. These insects use pheromones, volatile chemicals produced by members of the same species, to communicate complex messages. The interaction of pheromone detection with tactile signals and visual cues creates a robust communication system that allows for coordinated colony behaviours (d'Ettorre and Moore, 2008; Ma and Krings, 2009; Kocher and Cocroft, 2019).

In summary, insects have evolved intricate anatomical adaptations to effectively sense and respond to chemical cues in their environment. Sensory structures, especially those found on antennae and mouthparts, play a crucial role in detecting a wide range of chemical cues. Additionally, the integration of chemosensory information with other sensory modalities enhances an insect's ability to navigate its ecological niche, find mates, locate resources and evade predators. These anatomical adaptations reflect the complex interplay between insect physiology and their chemical ecology.

3.9 INSECT OLFACTION AND PHYTOCHEMICALS

As mentioned earlier, insects rely heavily on olfaction, their sense of smell, for locating food sources and mates, particularly crucial for herbivorous insects in finding suitable plants to feed on (Bruce, Wadhams and Woodcock, 2005; Conchou et al., 2019). Plants, in turn, produce chemicals known as secondary metabolites, influencing insect behaviour and serving as a defence mechanism (War et al., 2012; Divekar et al., 2022). The complex interaction between insects and plants is essential for understanding ecological systems and coevolution. The olfactory system of insects undergoes several steps to understand phytochemicals released in the environment. The process involves the detection of chemicals, specific odour receptors, olfactory sensilla in the antennae housing these receptors, neural processing in the brain and subsequent behavioural responses based on the interpretation of odour signals (Andersson, Löfstedt and Newcomb, 2015; Schmidt and Benton, 2020). Insects have evolved to recognise specific phytochemicals, allowing them to locate preferred hosts while avoiding less suitable plants (Scriber, 2002; Bruce and Pickett, 2011). This intricate system enables insects to navigate and survive in their environments, offering insights for pest control and conservation.

In addition to olfaction, insects utilise pheromones as chemical messengers for crucial communication within their populations (Shorey, 1973, 2013). Various types of pheromones play distinct roles, including sex pheromones for mate attraction, aggregation pheromones for group formation, alarm pheromones for signalling danger and trail pheromones for organised movements (Ali and Morgan, 1990; Abd El-Ghany, 2020). The production, release and perception of pheromones involve a finely tuned orchestration, with insects possessing specialised sensory structures such as antennae and sensilla. The impact of pheromones on insect behaviour, reproduction and ecology is profound. Sex pheromones orchestrate mating rituals, aggregation pheromones facilitate collective benefits, alarm pheromones trigger rapid escape responses and trail pheromones enable coordinated foraging in social insects. These chemical cues not only regulate individual behaviours but also shape the dynamics of entire populations and ecosystems (Fleischer and Krieger, 2018). In conclusion, the intricate world of pheromone detection and communication in insects exemplifies how chemical cues, in tandem with olfaction, shape their behaviour and ecology. The classification of various pheromone types, the intricate process of production and detection and their profound impact collectively highlight the significance of chemical communication in the natural world.

3.10 PRACTICAL APPLICATIONS OF INSECT CHEMORECEPTION

The study of chemoreception in insects extends beyond its fundamental role in ecological interactions to practical applications with broad implications. Pheromones, chemical messengers used by insects for communication, are central to agricultural practices. Leveraging these pheromones has led to the development of environmentally friendly alternatives for pest management (Mbaluto et al., 2020). Techniques such as mass trapping and mating disruption, made possible by identifying and synthesising specific pheromones, effectively reduce pest populations. This not only minimises crop damage but also decreases reliance on conventional chemical pesticides, offering sustainable solutions for agriculture (Rizvi et al., 2021). Additionally, the impact of chemoreception in insect-mediated disease transmission is a crucial area of application. Disease vectors such as mosquitoes, guided by olfactory cues, become targets for control strategies (Zwiebel and Takken, 2004). Understanding the compounds that attract these vectors opens avenues for developing targeted traps and repellents, significantly contributing to disease control and prevention efforts. This intersection of chemoreception and disease control showcases the practical implications of understanding insect sensory mechanisms (Muema et al., 2017; Stica, Lobo and Moore, 2021).

Moreover, chemosensory research has triggered advancements in sensory technology. By mimicking the finely tuned chemoreceptive systems of insects, artificial olfactory devices and biosensors have been developed (Saha, 2022). These innovations, capable of detecting minute amounts of specific chemicals, find applications in diverse fields, from environmental monitoring to food safety and security. The crossroads of insect-inspired technology and chemoreception reveal the transformative potential of this knowledge in addressing real-world challenges (Lu and Liu, 2022). In conclusion, the applied aspects of insect chemoreception knowledge weave a narrative that spans agriculture, disease control, technology and more. From sustainable pest management in agriculture to targeted disease control measures and innovative sensory technologies, the intricate world of insect chemoreception continues to shape and enhance our approach to solving multifaceted challenges in the natural world.

3.11 CONCLUSION

The exploration of chemoreception in insects illuminates the pivotal role of chemical communication in shaping ecological interactions. The intricate chemosensory mechanisms underscore the complexity of insect-environment relationships, influencing behaviours essential for survival and reproduction, from locating mates to responding to potential threats. Ongoing research into the molecular processes governing insect chemoreception offers exciting prospects, with advances in molecular biology, neurophysiology and sensory technology, promising novel strategies for pest management, disease control and conservation. Beyond its scientific context, the study of chemoreception provides profound insights into the broader intricacies of nature, enhancing our understanding of ecological relationships and adaptations that have shaped life on Earth. This knowledge informs efforts to address challenges posed by insect-mediated disease transmission, agricultural sustainability and biodiversity conservation. In summary, chemoreception in insects serves as a testament to the remarkable ways organisms have evolved to interact with their environments, with ongoing research holding the promise of unveiling further mysteries and innovations that continue to shape our understanding of nature's intricate web. Delving into the chemical conversations that shape the lives of insects enriches our understanding of the complex tapestry of life itself.

REFERENCES

Abd El-Ghany, N. M. (2020) 'Pheromones and chemical communication in insects', *Pests, weeds and diseases in agricultural crop and animal husbandry production*. IntechOpen, pp. 16–30.

Acevedo Ramos, F. *et al.* (2020) 'Comparative study of sensilla and other tegumentary structures of Myrmeleontidae larvae (Insecta, Neuroptera)', *Journal of Morphology*, 281(10), pp. 1191–1209.

Ahmad, F., Aslam, M. and Razaq, M. (2004) 'Chemical ecology of insects and tritrophic interactions', *Journal of Research (Sci.)*, 15(January 2004), pp. 181–190. Available at: http://www.bzu.edu.pk/jrscience/vol15no2/7.pdf.

Ali, M. F. and Morgan, E. D. (1990) 'Chemical communication in insect communities: a guide to insect pheromones with special emphasis on social insects', *Biological Reviews*, 65(3), pp. 227–247.

Ali, S. A. I. *et al.* (2015) 'Understanding insect behaviors and olfactory signal transduction', *Enliven: Journal of Genetic, Molecular and Cellular Biology*, 2(004).

Amer, A. and Mehlhorn, H. (2006) 'The sensilla of Aedes and Anopheles mosquitoes and their importance in repellency', *Parasitology Research*, 99, pp. 491–499.

Amrein, H. and Thorne, N. (2005) 'Gustatory perception and behavior in Drosophila melanogaster', *Current Biology*, 15(17), pp. R673–R684.

Andersson, M. N., Löfstedt, C. and Newcomb, R. D. (2015) 'Insect olfaction and the evolution of receptor tuning', *Frontiers in Ecology and Evolution*, 3, p. 53.

Anton, S. and Homberg, U. (1999) 'Antennal lobe structure', *Insect olfaction*, pp. 97–124.

Auer, T. O. *et al.* (2022) 'Copy number changes in co-expressed odorant receptor genes enable selection for sensory differences in drosophilid species', *Nature Ecology & Evolution*, 6(9), pp. 1343–1353.

Baeckens, S. (2019) 'Evolution of animal chemical communication: insights from non-model species and phylogenetic comparative methods', *Belgian Journal of Zoology/Koninklijke Belgische Vereniging voor Dierkunde.-Brussel, 1990, currens*, 149(1), pp. 63–93.

Barrozo, R. B. (2019) 'Food recognition in hematophagous insects', *Current Opinion in Insect Science*, 34, pp. 55–60.

Behmer, S. T. and Joern, A. (2012) 'Insect herbivore outbreaks viewed through a physiological framework: insights from Orthoptera', *Insect outbreaks revisited*, pp. 1–29.

Bell, W. J. and Cardé, R. T. (2013) *Chemical ecology of insects*. Springer.

Boeckh, J., Kaissling, K.-E. and Schneider, D. (1965) 'Insect olfactory receptors', *Cold Spring Harbor symposia on quantitative biology*. Cold Spring Harbor Laboratory Press, pp. 263–280.

Borkakati, R. N. *et al.* (2019) 'A brief review on food recognition by insects: use of sensory and behavioural mechanisms', *Journal of Entomology and Zoology Studies*, 7, pp. 574–579.

Böröczky, K. *et al.* (2013) 'Insects groom their antennae to enhance olfactory acuity', *Proceedings of the National Academy of Sciences*, 110(9), pp. 3615–3620.

de Brito Sanchez, M. G. (2011) 'Taste perception in honey bees', *Chemical Senses*, 36(8), pp. 675–692.

Bruce, T. J. A. and Pickett, J. A. (2011) 'Perception of plant volatile blends by herbivorous insects-finding the right mix', *Phytochemistry*, 72(13), pp. 1605–1611. doi: 10.1016/j.phytochem.2011.04.011.

Bruce, T. J. A., Wadhams, L. J. and Woodcock, C. M. (2005) 'Insect host location: a volatile situation', *Trends in Plant Science*, 10(6), pp. 269–274.

Bruner-Montero, G. *et al.* (2021) 'Symbiont-mediated protection of Acromyrmex Leaf-Cutter Ants from the entomopathogenic fungus Metarhizium anisopliae', *Mbio*, 12(6), pp. e01885–21.

Callahan, P. S. (1975) 'Insect antennae with special reference to the mechanism of scent detection and the evolution of the sensilla', *International Journal of Insect Morphology and Embryology*, 4(5), pp. 381–430.

Cardé, R. T. and Bell, W. J. (1995) *Chemical ecology of insects 2*. Springer Science & Business Media.

Cassau, S. *et al.* (2022) 'The sensilla-specific expression and subcellular localization of SNMP1 and SNMP2 reveal novel insights into their roles in the antenna of the desert locust Schistocerca gregaria', *Insects*, 13(7), p. 579.

Chakraborty, A. and Roy, A. (2021) 'Microbial influence on plant–insect interaction', *Plant-pest interactions: from molecular mechanisms to chemical ecology: chemical ecology*, pp. 337–363.

Chapman, R. F. (2003) 'Contact chemoreception in feeding by phytophagous insects', *Annual Review of Entomology*, 48(1), pp. 455–484.

Conchou, L. *et al.* (2019) 'Insect odorscapes: from plant volatiles to natural olfactory scenes', *Frontiers in physiology*, p. 972.

Crumière, A. J. J. *et al.* (2020) 'Using nutritional geometry to explore how social insects navigate nutritional landscapes', *Insects*, 11(1), p. 53.

d'Ettorre, P. and Moore, A. J. (2008) 'Chemical communication and the coordination of social interactions in insects', *Sociobiology of communication: an interdisciplinary perspective*, pp. 81–96.

Delompré, T. *et al.* (2019) 'Taste perception of nutrients found in nutritional supplements: a review', *Nutrients*, 11(9), p. 2050.

Depetris-Chauvin, A., Galagovsky, D. and Grosjean, Y. (2015) 'Chemicals and chemoreceptors: ecologically relevant signals driving behavior in Drosophila', *Frontiers in Ecology and Evolution*, 3, p. 41.

Dethier, V. G. (1957) 'Chemoreception and the behavior of insects', *Survey of biological progress*. Elsevier, pp. 149–183.

Dethier, V. G. and Chadwick, L. E. (1948) 'Chemoreception in insects', *Physiological Reviews*, 28(2), pp. 220–254.

Divekar, P. A. *et al.* (2022) 'Plant secondary metabolites as defense tools against herbivores for sustainable crop protection', *International Journal of Molecular Sciences*, 23(5), p. 2690.

Douglas, A. E. (2003) 'The nutritional physiology of aphids', *Advances in Insect Physiology*, 31(31), pp. 73–140.

Eisthen, H. L. (2002) 'Why are olfactory systems of different animals so similar?', *Brain, Behavior and Evolution*, 59(5–6), pp. 273–293.

Engl, T. and Kaltenpoth, M. (2018) 'Influence of microbial symbionts on insect pheromones', *Natural Product Reports*, 35(5), pp. 386–397.

Faucheux, M. J., Kristensen, N. P. and Yen, S.-H. (2006) 'The antennae of neopseustid moths: morphology and phylogenetic implications, with special reference to the sensilla (Insecta, Lepidoptera, Neopseustidae)', *Zoologischer Anzeiger-A Journal of Comparative Zoology*, 245(2), pp. 131–142.

Feng, Y. J. *et al.* (2015) 'Effect of Paranosema locustae (Microsporidia) on the behavioural phases of Locusta migratoria (Orthoptera: Acrididae) in the laboratory', *Biocontrol Science and Technology*, 25(1), pp. 48–55.

Ferrari, M. C. O., Wisenden, B. D. and Chivers, D. P. (2010) 'Chemical ecology of predator–prey interactions in aquatic ecosystems: a review and prospectus', *Canadian Journal of Zoology*, 88(7), pp. 698–724.

Fialho, M. do C. Q. *et al.* (2014) 'A comparative study of the antennal sensilla in corbiculate bees', *Journal of Apicultural Research*, 53(3), pp. 392–403.

Fleischer, J. and Krieger, J. (2018) 'Insect pheromone receptors–key elements in sensing intraspecific chemical signals', *Frontiers in Cellular Neuroscience*, 12, p. 425.

Freeman, E. G., Wisotsky, Z. and Dahanukar, A. (2014) 'Detection of sweet tastants by a conserved group of insect gustatory receptors', *Proceedings of the National Academy of Sciences*, 111(4), pp. 1598–1603.

Frings, H. and Frings, M. (1949) 'The loci of contact chemoreceptors in insects. A review with new evidence', *American Midland Naturalist*, 41, pp. 602–658.

Gainett, G. *et al.* (2017) 'Ultrastructure of chemoreceptive tarsal sensilla in an armored harvestman and evidence of olfaction across Laniatores (Arachnida, Opiliones)', *Arthropod Structure & Development*, 46(2), pp. 178–195.

Giorgini, M. *et al.* (2023) 'The susceptibility of Bemisia tabaci Mediterranean (MED) species to attack by a parasitoid wasp changes between two whitefly strains with different facultative endosymbiotic bacteria', *Insects*, 14(10), p. 808.

Hallem, E. A., Dahanukar, A. and Carlson, J. R. (2006) 'Insect odor and taste receptors', *Annual Review of Entomology*, 51, pp. 113–135.

Hansson, B. S. and Stensmyr, M. C. (2011) 'Evolution of insect olfaction', *Neuron*, 72(5), pp. 698–711.

Hansson, B. and Wicher, D. (2016) 'Chemical ecology in insects', *Chemosensory transduction*, pp. 29–45.

Heine, D. *et al.* (2018) 'Chemical warfare between leafcutter ant symbionts and a co-evolved pathogen', *Nature Communications*, 9(1), p. 2208.

Hu, P. *et al.* (2018) 'Sensilla on six olfactory organs of male Eogystia hippophaecolus (Lepidoptera: Cossidae)', *Microscopy Research and Technique*, 81(9), pp. 1059–1070.

Joseph, R. M. and Carlson, J. R. (2015) 'Drosophila chemoreceptors: a molecular interface between the chemical world and the brain', *Trends in Genetics*, 31(12), pp. 683–695.

Kaupp, U. B. (2010) 'Olfactory signalling in vertebrates and insects: differences and commonalities', *Nature Reviews Neuroscience*, 11(3), pp. 188–200.

King, B. H. and Gunathunga, P. B. (2023) 'Gustation in insects: taste qualities and types of evidence used to show taste function of specific body parts', *Journal of Insect Science*, 23(2), p. 11.

Kocher, S. D. and Cocroft, R. B. (2019) 'Signals in insect social organization', *Encyclopedia of animal behavior*. Elsevier, pp. 558–567.

Lee, J. K. and Strausfeld, N. J. (1990) 'Structure, distribution and number of surface sensilla and their receptor cells on the olfactory appendage of the male moth Manduca sexta', *Journal of Neurocytology*, 19(4), pp. 519–538.

Li, Q. and Montell, C. (2021) 'Mechanism for food texture preference based on grittiness', *Current Biology*, 31(9), pp. 1850–1861.

Lu, Y. and Liu, Q. (2022) 'Insect olfactory system inspired biosensors for odorant detection', *Sensors & Diagnostics*, 1, pp. 1126–1142.

Lucas, P., Montagné, N. and Jacquin-Joly, E. (2022) 'Anatomy and functioning of the insect chemosensory system', *Extended biocontrol*. Springer, pp. 183–195.

Ma, Z. S. and Krings, A. W. (2009) 'Insect sensory systems inspired computing and communications', *Ad Hoc Networks*, 7(4), pp. 742–755.

Mathis, A. and Crane, A. L. (2017) 'Chemoreception' in J. Call, G. M. Burghardt, I. M. Pepperberg, C. T. Snowdon and T. Zentall (Eds.), *APA handbook of comparative psychology: Perception, learning, and cognition* (pp. 69–87). American Psychological Association. https://doi.org/10.1037/0000012-00

Mbaluto, C. M. *et al.* (2020) 'Insect chemical ecology: chemically mediated interactions and novel applications in agriculture', *Arthropod-plant Interactions*, 14, pp. 671–684.

Melvin, R. G. *et al.* (2018) 'Natural variation in sugar tolerance associates with changes in signaling and mitochondrial ribosome biogenesis', *Elife*, 7, p. e40841.

Minnich, D. E. (1929) 'The chemical senses of insects', *The Quarterly Review of Biology*, 4(1), pp. 100–112.

Missbach, C. *et al.* (2014) 'Evolution of insect olfactory receptors', *elife*, 3, p. e02115.

Missbach, C. *et al.* (2020) 'Developmental and sexual divergence in the olfactory system of the marine insect Clunio marinus', *Scientific Reports*, 10(1), p. 2125.

Mitchell, B. K. (2009) 'Chemoreception', *Encyclopedia of insects*. Elsevier, pp. 148–152.

Mondal, S. *et al.* (2023) 'Insect microbial symbionts: ecology, interactions, and biological significance', *Microorganisms*, 11(11), p. 2665.

Moyano, A., Croce, A. C. and Scolari, F. (2023) 'Pathogen-mediated alterations of insect chemical communication: from pheromones to behavior', *Pathogens*, 12(11), p. 1350.

Muema, J. M. *et al.* (2017) 'Prospects for malaria control through manipulation of mosquito larval habitats and olfactory-mediated behavioural responses using plant-derived compounds', *Parasites & Vectors*, 10(1), pp. 1–18.

Murdoch, W. W., Evans, F. C. and Peterson, C. H. (1972) 'Diversity and pattern in plants and insects', *Ecology*, 53(5), pp. 819–829.

Nakanishi, A. *et al.* (2009) 'Sex-specific antennal sensory system in the ant Camponotus japonicus: structure and distribution of sensilla on the flagellum', *Cell and Tissue Research*, 338(1), pp. 79–97.

Noman, A. *et al.* (2020) 'Plant-insect-microbe interaction: a love triangle between enemies in ecosystem', *Science of the Total Environment*, 699, p. 134181.

Ozaki, M. and Wada-Katsumata, A. (2010) 'Perception and olfaction of cuticular compounds', *Insect hydrocarbons*, pp. 207–221.

Param, A. M. *et al.* (2022) 'Cuticular sensilla on the larval mouthparts of Antlion Myrmeleon sp. and its role in predatory behavior', *Proceedings of the National Academy of Sciences, India Section B: Biological Sciences*, 92(2), pp. 385–392.

Patt, J. M. *et al.* (2014) 'Innate and conditioned responses to chemosensory and visual cues in Asian citrus psyllid, Diaphorina citri (Hemiptera: Liviidae), vector of Huanglongbing pathogens', *Insects*, 5(4), pp. 921–941.

Rani, A. T. *et al.* (2021) 'Morphological characterization of antennal sensilla of Earias vittella (Fabricius)(Lepidoptera: Nolidae)', *Micron*, 140, p. 102957.

Ren, C.-S. *et al.* (2023) 'Comparison of morphological characteristics of Antennae and Antennal Sensilla among four species of Bumblebees (Hymenoptera: Apidae)', *Insects*, 14(3), p. 232.

Rizvi, S. A. H. *et al.* (2021) 'Latest developments in insect sex pheromone research and its application in agricultural pest management', *Insects*, 12(6), p. 484.

Rogers, S. M. and Newland, P. L. (2003) 'The neurobiology of taste in insects', *Advances in Insect Physiology*, 31, pp. 141–204.

Roy, H. E., Pell, J. K. and Alderson, P. G. (1999) 'Effects of fungal infection on the alarm response of pea aphids', *Journal of Invertebrate Pathology*, 74(1), pp. 69–75.

Ruedenauer, F. A. *et al.* (2023) 'The ecology of nutrient sensation and perception in insects', *Trends in Ecology & Evolution*, 38(10), pp. 994–1004.

Ryan, M. F. (2002a) *Insect chemoreception*. Springer.

Ryan, M. F. (2002b) 'Plant chemicals', *Insect chemoreception: fundamental and applied*, pp. 27–72.

Ryan, M. F. (2002c) 'The chemoreceptive organs: structural aspects', *Insect chemoreception: fundamental and applied*, pp. 113–139.

Sachse, S. and Krieger, J. (2011) 'Olfaction in insects: the primary processes of odor recognition and coding', *e-Neuroforum*, 2, pp. 49–60.

Saha, D. (2022) 'Insect olfaction in chemical sensing', *Canines: the original biosensors*. Jenny Stanford Publishing, pp. 151–177.

Santos-Garcia, D. *et al.* (2020) 'Inside out: microbiota dynamics during host-plant adaptation of whiteflies', *The ISME Journal*, 14(3), pp. 847–856.

Schmidt, H. R. and Benton, R. (2020) 'Molecular mechanisms of olfactory detection in insects: beyond receptors', *Open Biology*, 10(10), p. 200252.

Schneider, D. (1964) 'Insect antennae', *Annual Review of Entomology*, 9(1), pp. 103–122.

Schneider, R. W. S., Price, B. A. and Moore, P. A. (1998) 'Antennal morphology as a physical filter of olfaction: temporal tuning of the antennae of the honeybee, Apis mellifera', *Journal of Insect Physiology*, 44(7–8), pp. 677–684.

Schoonhoven, L. M. (1987) 'What makes a caterpillar eat? The sensory code underlying feeding behavior', *Perspectives in chemoreception and behavior*. Springer, pp. 69–97.

Schoonhoven, L. M. (2018) 'Insects in a chemical world', *Handbook of Natural Pesticides*, 6, pp. 1–21.

Schroeder, T. B. H. *et al.* (2018) 'It's not a bug, it's a feature: functional materials in insects', *Advanced Materials*, 30(19), p. 1705322.

Schuman, M. C. (2023) 'Where, when, and why do plant volatiles mediate ecological signaling? The answer is blowing in the wind', *Annual Review of Plant Biology*, 74, pp. 609–633.

Scriber, J. M. (2002) 'Evolution of insect-plant relationships: chemical constraints, coadaptation, and concordance of insect/plant traits', *Entomologia experimentalis et applicata*, 104(1), pp. 217–235.

Shields, V. D. C. and Shields, V. D. (2021) 'Functional morphology of gustatory organs in caterpillars', *Moths and caterpillars*. InTechOpen, p. 81.

Shorey, H. H. (1973) 'Behavioral responses to insect pheromones', *Annual Review of Entomology*, 18(1), pp. 349–380.

Shorey, H. H. (2013) *Animal communication by pheromones*. Academic Press.

Simcock, N. K. (2015) 'The expression and function of gustatory receptors in the honeybee (Apis mellifera)'. Newcastle Univerity.

Sokolinskaya, E. L. *et al.* (2020) 'Molecular principles of insect chemoreception', *Acta Naturae (англоязычная версия)*, 12(3(46)), pp. 81–91.

Song, L.-M. *et al.* (2017) 'Ultrastructure and morphology of antennal sensilla of the adult diving beetle Cybister japonicus Sharp', *PLoS One*, 12(3), p. e0174643.

Steinbrecht, R. A. (2007) 'Structure and function of insect olfactory sensilla', *Ciba Foundation Symposium 200-Olfaction in Mosquito-Host Interactions: Olfaction in Mosquito-Host Interactions: Ciba Foundation Symposium 200*. Wiley Online Library, pp. 158–183.

Stica, C., Lobo, N. F. and Moore, S. J. (2021) 'Peri-domestic vector control interventions using attractive targeted sugar baits and push-pull strategies', *Ecology and control of vector-borne diseases*. Wageningen Academic Publishers, pp. 645–652.

Thiagarajan, D. and Sachse, S. (2022) 'Multimodal information processing and associative learning in the insect brain', *Insects*, 13(4), p. 332.

Treanore, E., Derstine, N. and Amsalem, E. (2021) 'What can mechanisms underlying derived traits tell us about the evolution of social behavior?', *Annals of the Entomological Society of America*, 114(5), pp. 547–561.

Tucker, D. and Smith, J. C. (1969) 'The chemical senses', *Annual Review of Psychology*, 20(1), pp. 129–158.

Wang, J. *et al.* (2022) 'Morphology and distribution of Antennal Sensilla in an egg parasitoid wasp, Anastatus disparis (Hymenoptera: Eupelmidae)', *Journal of Insect Science*, 22(6), p. 6.

Wang, Y. and Dai, W. (2017) 'Fine structure of mouthparts and feeding performance of Pyrrhocoris sibiricus Kuschakevich with remarks on the specialization of sensilla and stylets for seed feeding', *PLoS One*, 12(5), p. e0177209.

Wang, Y., Li, L. and Dai, W. (2019) 'Fine morphology of the mouthparts in Cheilocapsus nigrescens (Hemiptera: Heteroptera: Miridae) reflects adaptation for phytophagous habits', *Insects*, 10(5), p. 143.

Wang, Z. et al. (2023) 'Influences of microbial symbionts on chemoreception of their insect hosts', *Insects*, 14(7), p. 638.

War, A. R. et al. (2012) 'Mechanisms of plant defense against insect herbivores', *Plant Signaling and Behavior*, 7(10). doi: 10.4161/psb.21663.

Wendin, K. M. E. and Nyberg, M. E. (2021) 'Factors influencing consumer perception and acceptability of insect-based foods', *Current Opinion in Food Science*, 40, pp. 67–71.

Wessnitzer, J. and Webb, B. (2006) 'Multimodal sensory integration in insects—towards insect brain control architectures', *Bioinspiration & Biomimetics*, 1(3), p. 63.

Wicher, D. and Miazzi, F. (2021) 'Functional properties of insect olfactory receptors: ionotropic receptors and odorant receptors', *Cell and Tissue Research*, 383, pp. 7–19.

Wikantyoso, B. et al. (2022) 'Ultrastructure and distribution of sensory receptors on the nonolfactory organs of the soldier caste in subterranean termite (Coptotermes spp.)', *Arthropod Structure & Development*, 70, p. 101201.

Wilson, E. O. (1965) 'Chemical communication in the social insects: insect societies are organized principally by complex systems of chemical signals', *Science*, 149(3688), pp. 1064–1071.

Xu, W. (2020) 'How do moth and butterfly taste?—Molecular basis of gustatory receptors in Lepidoptera', *Insect Science*, 27(6), pp. 1148–1157.

Xu, W., Zhang, H.-J. and Anderson, A. (2012) 'A sugar gustatory receptor identified from the foregut of cotton bollworm Helicoverpa armigera', *Journal of Chemical Ecology*, 38(12), pp. 1513–1520.

Zacharuk, R. Y. (1980) 'Ultrastructure and function of insect chemosensilla', *Annual Review of Entomology*, 25(1), pp. 27–47.

Zahran, N., Sawires, S. and Hamza, A. (2022) 'Piercing and sucking mouth parts sensilla of irradiated mosquito, Culex pipiens (Diptera: Culicidae) with gamma radiation', *Scientific Reports*, 12(1), p. 17833.

Zhang, W. et al. (2016) 'Tissue, developmental, and caste-specific expression of odorant binding proteins in a eusocial insect, the red imported fire ant, Solenopsis invicta', *Scientific Reports*, 6(1), p. 35452.

Zu, P. et al. (2023) 'Plant–insect chemical communication in ecological communities: an information theory perspective', *Journal of Systematics and Evolution*, 61(3), pp. 445–453.

Zwiebel, L. J. and Takken, W. (2004) 'Olfactory regulation of mosquito–host interactions', *Insect Biochemistry and Molecular Biology*, 34(7), pp. 645–652.

CHAPTER 4

Plant Chemical Defence Against Insect Herbivores

4.1 INTRODUCTION

Plant–insect interactions have played a pivotal role in shaping terrestrial ecosystems throughout evolutionary history. These interactions are not merely casual encounters; rather, they represent complex and dynamic relationships that have driven the coevolution of both plant defences and insect adaptations. The fundamental aspect of these exchanges centres on the complex realm of chemical defences, wherein plants have evolved a remarkable assortment of molecular armaments to counteract the insatiable feeding behaviours of herbivorous insects (Mithöfer, Boland and Maffei, 2009). Chemical defences are the cornerstone of plant–insect interactions, serving as the frontline guardians of plants against the relentless onslaught of herbivores (War et al., 2012). Insects, representing one of the most diverse and ubiquitous groups of organisms on Earth, have evolved a myriad of feeding strategies, behaviours and adaptations to exploit plant resources (War et al., 2018). In response, plants have evolved equally diverse and ingenious strategies to fend off herbivores. Chemical defences emerge as a fundamental component of these strategies, offering a means of deterring, repelling or even incapacitating herbivores (Bruce, 2015).

DOI: 10.1201/9781003479857-4

Over millions of years, plants have embarked on an evolutionary arms race with herbivores, resulting in the development of sophisticated chemical defence mechanisms. These mechanisms involve the synthesis and deployment of an astonishing array of secondary metabolites, ranging from alkaloids and terpenoids to phenolics and glucosinolates (War et al., 2018). Each of these compounds serves as a strategic tool in the plant's arsenal, aimed at thwarting herbivore attacks and minimising the fitness costs associated with such interactions. The study of plant chemical defence responses against insect herbivores holds profound implications for both agricultural practices and ecological dynamics (Ahuja, Rohloff and Bones, 2010; James et al., 2012). With the global population soaring and the need for sustainable food production becoming increasingly urgent, the efficient management of pest herbivores is of paramount importance (James et al., 2012). Understanding how plants synthesise and deploy defensive compounds can pave the way for the development of novel and eco-friendly pest management strategies that minimise the reliance on chemical pesticides.

Furthermore, the complex web of interactions between plants, herbivores and their natural enemies orchestrates the delicate balance of ecosystems (Vucetic et al., 2014). By deciphering the complex chemical dialogues that underpin these interactions, we gain insights into the broader ecological dynamics that shape community structures and trophic relationships (Agrawal, 2011). The revelation of how certain plants manipulate herbivores through chemical cues, or how herbivores adapt to circumvent plant defences, unveils a world of natural strategies that could be harnessed for conservation and restoration efforts. In this chapter, we embark on a journey into the fascinating realm of plant chemical defence responses against herbivorous insects. We delve into the intricate molecular interactions that govern the synthesis, perception and deployment of defensive compounds. As we navigate through the mechanisms of plant–herbivore coevolution, the ecological implications of chemical defences become apparent. From pest management to the preservation of delicate ecosystems, the understanding of plant chemical defences shines as a beacon of insight, guiding us toward sustainable and harmonious interactions between plants and their insect adversaries.

4.2 CHEMICAL CONSTITUENTS OF PLANT DEFENCES

Plants, tethered to the ground and besieged by various biotic and abiotic stresses, have evolved a sophisticated array of defence mechanisms to combat threats. Biotic stress, primarily posed by heterotrophic organisms seeking to exploit plant energy, requires plants to deploy diverse strategies. Central to these strategies is the intricate chemistry of plants, boasting more than 200,000 specialised metabolites (Axel Mithöfer and Boland, 2012). These chemical defences, play a crucial role in shaping plant–herbivore interactions and influencing the broader ecology of ecosystems (Erb and Kliebenstein, 2020; Divekar et al., 2022). Plants have developed an intricate repertoire of chemical compounds as a direct response to the evolutionary pressure exerted by herbivores. These compounds, often referred to as secondary metabolites, are not central to the plant's primary metabolic pathways but are, instead, synthesised in specific tissues and under particular conditions, often in response to herbivore attacks (Bennett and Wallsgrove, 1994; Crozier, Clifford and Ashihara, 2006). The resulting chemical diversity equips plants with a versatile toolkit to combat a wide array of herbivores and their feeding strategies.

The secondary metabolites encompass an extensive spectrum of chemical classes, each with its unique properties and modes of action. Alkaloids, for instance, are nitrogen-containing compounds that can act as potent toxins when ingested by herbivores (Matsuura and Fett-Neto, 2015). Terpenoids, on the other hand, constitute a diverse group of compounds with a broad range of functions, from deterring herbivores through their strong aromas to acting as direct defences against them (Gershenzon and Dudareva, 2007). Phenolics, another class of secondary metabolites, include compounds such as tannins and lignins (Crozier, Jaganath and Clifford, 2006; Lattanzio et al., 2009; Hassanpour, MaheriSis and Eshratkhah, 2011). Tannins are known for their astringent taste and have the capacity to bind with proteins, interfering with herbivores' digestion and nutrient absorption (Hassanpour, MaheriSis and Eshratkhah, 2011). Lignins contribute to the structural integrity of plant tissues and can hinder herbivore feeding (Jung and Allen, 1995; Frei, 2013; Fürstenberg-Hägg, Zagrobelny and Bak, 2013). Glucosinolates, prevalent in members of the Brassicaceae family, are sulphur-containing compounds with a dual role (Halkier and Gershenzon, 2006). While they act as feeding deterrents, their real potency emerges upon tissue damage when they are enzymatically converted into toxic isothiocyanates, thwarting herbivores' attempts to consume the plant (Bruce, 2014; Singh, 2017).

The diverse secondary metabolites serve various functions in the plant's defensive strategy. Some compounds act as powerful deterrents, discouraging herbivores from feeding due to their unpalatable taste or toxic effects

(Chen, 2008). Others function as direct toxins, incapacitating herbivores upon ingestion (Axel Mithöfer and Boland, 2012). Moreover, some plants have harnessed the power of chemical mimicry, releasing volatile compounds that attract herbivore natural enemies, such as parasitoids and predators, to aid in their protection (Ali et al., 2021, 2023). Understanding the complex relationship between plants and their herbivore adversaries requires a deep appreciation of the multifaceted roles these chemical constituents play. The evolution of these compounds reflects not only the plants' ongoing battle for survival but also the interconnectedness of species in complex ecological webs. In this exploration of the chemical constituents of plant defences, we have skimmed the surface of the diverse and intricate mechanisms that plants employ to deter, thwart and manipulate herbivores. From alkaloids to glucosinolates, the chemistry of these compounds intersects with the ecology of plant–insect interactions in ways that continue to captivate scientists and hold promising applications for sustainable agriculture and ecosystem management.

The chemical defence mechanisms employed by plants against herbivorous insects are categorised into two primary modes: direct and indirect defences, each delineated by their distinct efficacies in managing herbivorous insects (Axel Mithöfer and Boland, 2012; War et al., 2012). Within the context of direct chemical defence, we scrutinise chemical compounds that exert a direct influence on the performance and behaviour of pests.

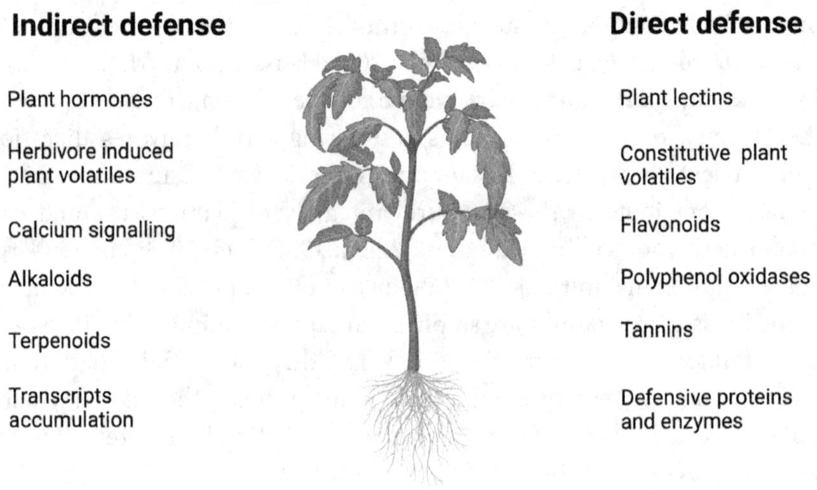

Indirect defense

Plant hormones

Herbivore induced
plant volatiles

Calcium signalling

Alkaloids

Terpenoids

Transcripts
accumulation

Direct defense

Plant lectins

Constitutive plant
volatiles

Flavonoids

Polyphenol oxidases

Tannins

Defensive proteins
and enzymes

FIGURE 4.1   Illustration showing chemical defence employed by plants against insect herbivores.

Conversely, indirect chemical defences encompass compounds that, while not directly impacting pest performance, enhance the recruitment of bio-control agents. It is imperative to acknowledge the inherent complexity of this classification, as certain chemical compounds may exhibit dual functionality, influencing herbivore behaviour both directly and indirectly. Moreover, the effectiveness of these chemical defences is contingent upon a myriad of factors, including but not limited to host plant status, environmental conditions and the specific interactions between plant species and herbivorous insects. This nuanced interplay underscores the intricate nature of plant–insect interactions, necessitating a comprehensive understanding of the multifaceted dynamics at play.

4.3 EVOLUTION OF CHEMICAL DEFENCES

Over millions of years, plants, insects and their predators have engaged in a chemical arms race, showcasing the refinement of chemical defence systems in response to the evolving strategies of antagonists. While widely accepted, empirical evidence supporting this concept is limited, with notable exceptions such as the evolution of angular furanocoumarins following the adaptation to linear furanocoumarins (Berenbaum and Zangerl, 1998). Among interactions between plants and herbivores, specialist herbivores demonstrate a remarkable resilience against host plant chemical defences compared to their generalist counterparts (Ali and Agrawal, 2012). Evolutionary adaptations manifest through mechanisms that detoxify, sequester, excrete or selectively bind plant defence compounds. The larvae of *Pieris rapae*, specialised in Brassicaceae plants, exhibit gut proteins directing hydrolysis toward nitriles, bypassing the toxic isothiocyanates (Wittstock et al., 2004; Ali, Tonğa, et al., 2024). Similarly, the diamondback moth, *Plutella xylostella*, employs a glucosinolate sulfatase to generate desulfoglucosinolates, outcompeting myrosinase and avoiding the production of toxic compounds (Ratzka et al., 2002). Behavioural adaptations, such as the diversification of cytochrome P450 monooxygenases in *Papilio polyxenes*, showcase evolutionary responses to feeding on toxin-containing host plants (Scott and Wen, 2001).

Chemical complexities in plants and insects often mirror an evolutionary arms race. In the Apiaceae plant, *Pastinaca sativa*, the production of furanocoumarins correlates with the herbivorous insect's ability to metabolise these compounds, suggesting a tightly linked genetic interplay (Berenbaum and Zangerl, 1998). Studies on the genus *Bursera* demonstrate a net accumulation of new compounds during species diversification, reinforcing the coevolution theory (A Mithöfer and Boland, 2012). Notably, insects may exploit plant-derived compounds for their own defence against parasitoids and predators, emphasising the intricate interplay between species (Beran and Petschenka, 2022; Wetzel et al., 2023). For instance, *P. xylostella* females are stimulated by glucosinolates to oviposit eggs on Brassicaceae (Renwick et al., 2006; Bruce, 2015). During insect–plant interaction, chemical compounds often confer a high level of resistance to specific plant hosts against insect pests. The success of herbivores in handling these compounds may, in turn, generate resistance. However, the perpetual arms race unfolds as each protagonist seeks to surpass the other, escalating the struggle between plant hosts and herbivores. As chemical defences evolve, so do the strategies employed by herbivores, creating a dynamic landscape of coevolution in the intricate theatre of nature (Birnbaum and Abbot, 2018; Yates and Michel, 2018; War et al., 2020).

4.4 PLANT CHEMICAL DEFENCE COMPOUNDS

4.4.1 Terpenes

Terpenes, a captivating class of plant secondary metabolites, stand as the largest and most diverse group among their botanical counterparts, boasting a staggering repertoire of around 25,000 compounds (Divekar et al., 2022). Derived from isoprene units, terpenes play a pivotal role in shaping the complex symphony of plant interactions, weaving a tapestry of chemical defences that extend beyond mere deterrence. The synthesis of terpenes is a finely tuned process, intricately regulated by organ, tissue, cell and developmental cues. Transcription factors choreograph this symphony, activating and transcribing the genes responsible for secondary metabolite production. Terpenes emerge from the marriage of isopentenyl diphosphate (IPP) and dimethylallyl diphosphate (DMAPP), synthesised through two distinct pathways: the cytosolic mevalonate (MVA) and plastidial 2-C-methyl-D-erythritol-4-phosphate (MEP) pathways (Ninkuu et al., 2021).

Classification further delineates the terpene realm into monoterpenes (C10), sesquiterpenes (C15), diterpenes (C20), triterpenes (C30) and tetraterpenes (C40), each with a distinct number of isoprene units (Croteau, Kutchan and Lewis, 2000; Mathur et al., 2024). Monoterpenes, with their two isoprene units, dance through the chloroplast, while sesquiterpenes find their rhythm in the endoplasmic reticulum or cytosol. Triterpenes, with their stately six isoprene units, showcase their elegance across diverse phytohormones, pigments and sterols. Terpenes don many hats in the botanical theatre, serving as essential components in phytohormones, pigments and sterols (Degenhardt, Köllner and Gershenzon, 2009). However, their true prowess lies in their role as allelochemicals, defensive toxins and herbivore deterrents (Paré and Tumlinson, 1999; Mumm, Posthumus and Dicke, 2008). Demonstrating a remarkable versatility, terpenes have been unveiled as potential plant defence regulators. Their lipophilic nature allows them to orchestrate a symphony with globular proteins, altering three-dimensional (3D) structures and playing a crucial role in plant defence strategies (Gershenzon and Dudareva, 2007).

Among terpenes, some standout performers, such as azadirachtin, emerge as insecticidal powerhouses; these compounds exhibit low toxicity to non-target organisms while effectively deterring insect adversaries (Gershenzon and Dudareva, 2007; Divekar et al., 2022). The triterpene azadirachtin, in particular, stands tall as a potent insect deterrent, showcasing the delicate balance between protection and ecological harmony. Terpenes, in their aromatic grandeur, exemplify nature's prowess in crafting intricate defence strategies. From repelling herbivores to beckoning natural enemies, they script a nuanced story in the botanical world, showcasing the perpetual dance of life in the green embrace of flora (Mumm, Posthumus and Dicke, 2008; Dambolena et al., 2016; Divekar et al., 2022).

4.4.2 Phenolic Compounds

Phenolic compounds, a diverse category of plant secondary metabolites, stand as a formidable force in the botanical world, comprising nearly 10,000 compounds (Divekar et al., 2022). Characterised by the presence of a hydroxyl functional group (phenyl group) on an aromatic ring, these compounds play a multifaceted role in protecting plants against herbivores and enticing pollinators (Lattanzio, 2013; War et al., 2020; Al Mamari, 2021). The chemical diversity within phenolics encompasses simple phenols such as catechols and hydroxybenzoic acid derivatives, along with flavonoids, catechol melanins, stilbenes, condensed tannins and lignins (Al Mamari, 2021). This chemical diversity equips plants with a versatile toolkit for defence mechanisms. Phenolics emerge as frontline defenders in the direct chemical warfare against herbivorous insects (Lattanzio, 2013; War et al., 2020). This group includes flavonoids, such as anthocyanins, and isoflavonoids, such as phytoalexins, medicarpin, rishitin and camalexin (Pratyusha, 2022). Flavonoids, acting as water-soluble pigments, shield plant foliage from ultraviolet radiation, while isoflavonoids exhibit antibiotic and antifungal properties in response to pathogen attacks (Falcone Ferreyra, Rius and Casati, 2012). Stored in vacuoles, tannins confer toxicity to insects by binding to salivary proteins and digestive enzymes, inhibiting essential protein functions. Lignin, a polymer in cell walls, constructs an indigestible physical barrier against herbivores (Simmonds, 2003; Rehman, Khan and Badruddin, 2012; Mierziak, Kostyn and Kulma, 2014).

The intricate biosynthetic pathway of phenolic compounds unfolds through the shikimate pathway, with shikimic acid as the precursor. This acid, derived from a combination of erythrose 4-phosphate and phosphoenolpyruvate, sets the stage for the synthesis of amino acids such as phenylalanine, tyrosine and tryptophan (Haslam, 2014; Santos-Sánchez et al., 2019). These amino acids, products of the shikimate pathway, serve as precursors for various phenolics, contributing to the plant's chemical defence arsenal. Constitutive defences in plants predominantly rely on carbon-based metabolites such as terpenes and polyphenols. Phenolic compounds exhibit their defensive prowess by interacting with target organism proteins, forming multiple hydrogen and ionic bonds that modify protein 3D structures (Divekar et al., 2022). The impact of these secondary metabolites is evident in instances such as the altered biological parameters of the maize stalk borer (*Chilo partellus*) on resistant maize germplasm, indicating an antibiosis effect induced by plant-produced secondary metabolites in response to larval feeding (Divekar, Kumar and Suby, 2019). Diet incorporation assays with phenolic compounds, such as ferulic and p-coumaric acids, further underscore their antibiosis effects on larvae, emphasising the intricate role these compounds play in shaping plant–insect interactions (Kumar and Suby, 2013).

4.4.3 Nitrogen-Containing Plant Secondary Compounds

In the botanical realm, a fascinating class of secondary metabolites takes centre stage: the nitrogen-containing secondary metabolites, primarily represented by alkaloids (Crozier, Clifford and Ashihara, 2006). This chemical saga unfolds across the vast landscape of the plant kingdom, in which approximately 10,000 different alkaloid derivatives have been documented to date (Isman and Paluch, 2011; Divekar et al., 2022). Alkaloids, categorised into three groups based on biosynthesis, showcase a remarkable diversity. True alkaloids, exemplified by nicotine, morphine, quinine and atropine, originate from amino acids, sharing a nitrogen-containing heterocyclic ring. Pseudo-alkaloids, including capsaicin, solanidine and caffeine, diverge from amino acid origins. Proto-alkaloids, such as yohimbine, mescaline and hordenine, derive nitrogen from amino acids but feature a distinct structural composition (Lichman, 2021; Bui, Rodríguez-López and Dang, 2023). Alkaloids further find themselves classified into various subcategories, including cyanogenic glycosides (CGs), glucosinolates, non-protein amino acids (NPAAs), aliphatic NPAAs, NPAAs with aromatic skeletons and NPAAs with cyclic and heterocyclic skeletons (Singh, 2018).

The multifaceted functions of alkaloids extend beyond their structural diversity. These compounds play a pivotal role in plant defence mechanisms, swiftly responding to pathogen attacks and adverse conditions (Koshiyama et al., 2006; Singh, 2018). Alkaloids act as deterrents against herbivores, disrupting neuronal signal transduction and interfering with crucial cellular processes such as DNA replication, protein synthesis and enzyme activity (Wink, Schmeller and Latz-Brüning, 1998). Their toxic and repellent effects serve as a sophisticated defence strategy against a spectrum of herbivores, aiming to reduce or prevent damage to plants (Matsuura and Fett-Neto, 2015). In the intricate dance of plant–insect interactions, alkaloids emerge as guardians, orchestrating a chemical symphony that underscores the resilience and adaptability of the botanical world (Macel, 2011; Wink, 2017). The biopathways leading to these nitrogen-containing secondary metabolites navigate through the tricarboxylic acid cycle and the shikimate pathway. Shikimic acid, the precursor, emerges from a combination of erythrose 4-phosphate (pentose phosphate pathway) and phosphoenolpyruvate (glycolytic pathway) (Divekar et al., 2022). Amino acids such as phenylalanine, tyrosine and tryptophan, products of the shikimate pathway, serve as crucial precursors not only for alkaloids but also for phenolics, weaving an intricate chemical tapestry within plants (Kumar et al., 2023).

4.4.4 Sulphur-Containing Plant Secondary Compounds

Sulphur-containing secondary metabolites, predominantly identified in the Brassicaceae and Capparales plant taxa, stand as stalwart defenders in the intricate dance of plant–insect interactions. With an array of approximately 120 reported molecular structures, these glucosides, known as glucosinolates, derive their uniqueness from amino acids (Divekar et al., 2022). The amino acid's side chain precursor determines the type of glucoside, with aliphatic glucosinolates originating from methionine, indole glucosinolates from tryptophan, aromatic glucosinolates from tyrosine or phenylalanine and an unidentified amino acid giving rise to another sub-group (Kitainda and Jez, 2021). Glucosinolates, typically housed in cell vacuoles and protected by myrosinase, play a pivotal role in plant defence (Ahuja, Rohloff and Bones, 2010; Lv et al., 2022). Herbivore-induced disruption triggers myrosinase to break down glucosinolates into toxic metabolites, such as nitriles, thiocyanates and isothiocyanates, mirror-ing the efficacy of synthetic insecticides and proving highly toxic to her-bivorous insects, acting both as deterrents and repellents from feeding (Agrawal and Kurashige, 2003; Axel Mithöfer and Boland, 2012; Chhajed et al., 2019).

The orchestration of sulphur uptake and biosynthesis pathways unfolds as plants absorb sulphate through roots, mainly in the form of sulphate (SO_4^{2-}), enhanced by metallic cations such as copper (Cu), selenium (Se) and zinc (Zn). The absorbed sulphate undergoes activation by con-densation with adenosine triphosphate (ATP), giving rise to adenosine 5'-phosphosulphate (APS) via ATP sulfurylase (Mithen, 2008; Abdalla and Mühling, 2019; Divekar et al., 2022). This initiates two crucial path-ways: the sulfation pathway, involving the phosphorylation of APS to form phosphoadenosine 5'-phosphosulphate (PAPS), responsible for sulphate group donation to molecules in primary and secondary metabolism, and the sulphate reductive pathway, leading to the biosynthesis of sulfolipids and sulphur-containing amino acids such as cysteine (Cys) and methio-nine (Met). Cys further contributes to the synthesis of glutathione, a vital mediator of cellular redox status and detoxification, while Met serves as the precursor of S-adenosylmethionine (SAM), a methyl group donor in various metabolic reactions (Divekar et al., 2022). Within this intricate realm, a diverse classification of sulphur-containing secondary metabo-lites, including glutathione, glucosinolates, phytoalexins, defensin and cysteine sulfoxides, emerges (Mithen, 2008). The biosynthetic journey intertwines with the tricarboxylic acid cycle and the shikimate pathway, with shikimic acid serving as a precursor. This pathway yields amino acids such as phenylalanine, tyrosine and tryptophan, not only for sulphur-con-taining metabolites but also for nitrogen-containing secondary metabo-lites and phenolics. In the grand symphony of plant defence, alkaloids, significant components produced by numerous higher plant species, play a pivotal role. Alkaloids, with their toxic and repellent effects, become ver-satile defenders against a broad spectrum of herbivores, disrupting cell signalling and showcasing the dynamic nature of plant chemical defences in navigating adverse conditions (Abdalla and Mühling, 2019; Divekar et al., 2022).

4.5 PLANT CHEMICAL DEFENCE BASED ON MODE OF ACTION

4.5.1 Direct Chemical Defence

Plant chemical compounds that directly affect pest performance and behaviour are considered direct defences. These chemical compounds influence pest behaviour, such as feeding, mating and ovipositional behaviour. These compounds may act as deterrents, antifeedants and antibiotics. Nitrogen compounds, exemplified by alkaloids, present another facet of direct chemical defence (Hadacek, 2002; Singh, 2018). Caffeine, nicotine, atropine and capsaicinoids exhibit toxicity against herbivores (Divekar et al., 2022). Phenolics constitute a crucial category of secondary metabolites in plants, playing a pivotal role in direct chemical defence against herbivorous insects. This diverse group includes flavonoids, such as anthocyanins and isoflavonoids such as phytoalexins, medicarpin, rishitin, camalexin, tannins, lignin and furanocoumarins (Lin et al., 2016; Pratyusha, 2022). Flavonoids, acting as water-soluble pigments, shield plant foliage from ultraviolet radiation, while isoflavonoids such as medicarpin, rishitin and camalexin exhibit antibiotic and antifungal properties in response to pathogen attack (Samanta, Das and Das, 2011). Stored in vacuoles, tannins confer toxicity to insects by binding to salivary proteins and digestive enzymes, thereby inhibiting protein function (Schultz, 1989; War et al., 2020). Lignin, forming a polymer in cell walls, establishes an indigestible physical barrier against herbivores (Moore and Jung, 2001; Uddin et al., 2024). CGs, including glucosinolates and specific compounds, exemplify unique modes of action against herbivores (Poulton, 1983; Selmar, 2010; Pentzold et al., 2015). CGs, stored as inactive conjugates, release hydrogen cyanide upon herbivore-induced cell damage, inhibiting cellular respiration. Protective mechanisms involve storage in vacuoles and the formation of a multienzyme complex, preventing the release of harmful intermediates during biosynthesis. Glucosinolates, compartmentalised within plant tissues, evade hydrolysis until tissue damage occurs. The resulting breakdown products, nitriles and isothiocyanates, exhibit toxicity against herbivores and influence host selection by insects (Hopkins, van Dam and van Loon, 2009).

Proteins and enzymes, such as proteinase inhibitors (PIs), play a crucial role in protecting proteins with antinutritional or toxic properties from

degradation by herbivore-induced proteolysis. PIs interfere with nutrient utilisation in the insect gut, contributing to plant resistance against
herbivores. The intricate interplay between PIs and defensive proteins
underscores the complexity of plant–insect interactions at the molecular
level (Divekar et al., 2023). Latex, a chemically diverse suspension found
in laticifers, serves as a defence mechanism rich in compounds such as
rubber, cardenolides and proteins. Latex acts by entangling herbivores,
making it an effective deterrent. Despite challenges in identifying the
precise active compounds within latex, its role in deterring herbivory is
well-established. From stickiness to the array of toxic components, latex
represents a multifaceted defence strategy employed by plants. While substantial progress has been made in understanding the broad categories of
plant defences and the roles of specific compounds, numerous questions
persist (Ramos et al., 2019).

4.5.2 Indirect Chemical Defence

In contrast to direct chemical defence, indirect strategies come to the fore.
Indirect chemical defence, primarily involving volatile organic compounds
(VOCs) and herbivore-induced plant volatiles, emerges as a sophisticated
approach in curtailing pest performance by actively engaging biocontrol
agents (Aljbory and Chen, 2018). Within the intricate dynamics of plant
defence against herbivorous insects, these mechanisms play a pivotal
role, orchestrating responses that extend beyond direct interactions. The
deployment of specific compounds, released into the plant's surroundings,
becomes a strategic manoeuvre shaping the behaviour and interactions of
both herbivores and their natural enemies. This exploration navigates the
distinctive landscape of indirect chemical defence, unravelling its complexities and illuminating its critical role in fortifying plant resilience.

A diverse array of volatile compounds, including alkaloids, terpenes,
terpenoids, nitrogen compounds, volatile indoles and fatty acid derivatives (green leaf volatiles), collectively form a sophisticated repertoire contributing significantly to indirect chemical defence in plants (Heil, 2008).
These volatiles act as airborne signals, signalling the presence of herbivores to nearby plants, thereby triggering the activation of defence mechanisms. The intricate arsenal also includes alcohols, aldehydes, benzenoids
and ketones released into the air during a plant's defence response, shaping the complex chemical cues that influence interactions between herbivores, plants and natural enemies in the ecosystem (Dicke and Hilker,
2003). Monoterpenes, homoterpenes and sesquiterpenes, categorised

as terpenoids, further enrich the indirect defence strategy. Released in response to herbivore feeding, these compounds not only serve as deterrents for herbivores but also attract natural enemies, contributing to a nuanced defence approach (Toffolatti et al., 2021). Terpenoids, derived from isoprene units, play a significant role in both direct and indirect defences, influencing herbivore behaviour, attracting parasitoids and impacting plant–plant interactions. The mechanisms by which terpenoids affect insect pests, potentially through alkylation of nucleophiles or interference with moulting regulation, are subjects of ongoing exploration (Mumm, Posthumus and Dicke, 2008; Pichersky and Raguso, 2018).

Jasmonic acid, salicylic acid and ethylene emerge as key players in the realm of indirect chemical defence. These plant hormones orchestrate complex signalling pathways that trigger defensive responses, influencing not only the attacked plant but also neighbouring vegetation. However, it is important to note that some of these chemical compounds also play a role as direct chemical defences (Tamaoki et al., 2013). Therefore, there is an overlap in their functions, sometimes acting as direct defences and sometimes as indirect defences depending upon the context and specificity of the insect and plant. Alkaloids, representing a structurally diverse group, serve as multifunctional defence compounds (Srivasatava, 2022). Nicotine, synthesised in response to herbivory, targets insect nicotinic acetylcholine receptors, while caffeine in *Coffea arabica* acts as a natural insecticide. Alkaloids exert versatile impacts on herbivores, affecting metabolic systems, enzymes and nervous systems, underscoring their significance in plant defence (Matsuura and Fett-Neto, 2015). This intricate network of compounds highlights the multifaceted strategies employed by plants to safeguard against herbivorous adversaries, ultimately contributing to the suppression of pest populations.

4.6 MODE OF PRODUCTION OF CHEMICAL DEFENCE IN PLANTS

Plant chemical defences are intricately regulated processes that involve the production of a diverse array of compounds aimed at deterring or repelling herbivores. The mode of production can be broadly categorised into constitutive and induced defences, each playing a crucial role in the plant's strategy to ward off herbivorous threats. A significant differentiation lies in the distinction between defence mechanisms that thwart full-blown infection from occurring in the initial stages (constitutive) and those that come into play after the onset of infection (induced) (Boots and

Best, 2018). Moreover, plants facing frequent or severe damage may find it advantageous to primarily allocate resources to constitutive defence, while those infrequently attacked may predominantly depend on induced defences (McKey, 1979).

4.6.1 Constitutive Chemical Defence

The term constitutive defence refers to chemical compounds continuously produced in plants, regardless of immediate herbivore presence (War et al., 2012). These constitutive chemical defences, including various toxins, serve as a consistent deterrent against herbivores and are a crucial component of the plant's defence mechanism (Wittstock and Gershenzon, 2002; Boots and Best, 2018). Involving toxic compounds, constitutive defences require plants to synthesise and store them without self-poisoning. One effective strategy is storing toxins as inactive precursors, such as glycosides, kept separate from activating enzymes (Jones and Vogt, 2001). For example, glucosinolates in capparales plants are compartmentalised away from their activating enzyme, myrosinase. Constitutive defence extends to the strategic deployment of chemical defences in plant parts deemed of high fitness value or at a high risk of attack. A field survey on wild parsnip (*Pastinaca sativa*) reveals a fascinating dynamic. Reproductive organs of this plant are frequently targeted by herbivores, prompting the accumulation of high constitutive levels of the toxic furanocoumarin, xanthotoxin, with no increase upon artificial damage. In contrast, the roots, rarely attacked, maintain low constitutive levels of xanthotoxin, which swiftly rise when wounded (Zangerl and Rutledge, 1996). In the realm of constitutive chemical defence, plants utilise glycosides such as glucosinolates and benzoxazinoids, primarily found in Gramineae. Tissue damage triggers the reversible hydrolysis of inactive D-glucoside precursors, producing phytotoxic aglycones such as 2,4-dihydroxy-1,4-benzoxazin-3-one (DIBOA) and its derivative DIMBOA. In maize, detoxification involves the reformation of D-glucosides catalysed by glucosyltransferases encoded by BX8 and BX9 genes (Von Rad et al., 2001; Ali, Mukarram, et al., 2024). Dicotyledonous species encountering benzoxazinoids from neighbouring grasses detoxify the DIBOA-decomposition product benzoxazolin-2-one (BOA) through hydroxylation and *N*-glycosylation (Wittstock and Gershenzon, 2002). This intricate orchestration exemplifies the multifaceted nature of constitutive chemical defence mechanisms in plants.

4.6.2 Induced Chemical Defences

Plants have evolved a sophisticated array of defence mechanisms, both constitutive and induced, to protect themselves from invading herbivores (Jander and Howe, 2008). Induced defence mechanisms are activated in response to specific stimuli, such as herbivore attacks, allowing plants to mount a rapid and targeted response (A Mithöfer and Boland, 2012). This dynamic interplay between constitutive and induced chemical defences equips plants with an adaptable strategy against herbivores, ensuring their survival in a dynamic environment. In the face of herbivore threats, plants rapidly detect attacks and trigger a cascade of defence mechanisms, allocating resources specifically to confront the imminent threat (Vallad and Goodman, 2004). Induced defences involve a complex interplay of biochemical pathways, resulting in the production of compounds that deter or inhibit herbivores. Notably, plants employ a sophisticated mechanism called priming, pre-activating defence pathways after an initial herbivore attack. This anticipatory response allows the plant to mount a quicker and more potent defence against subsequent herbivore attacks, enhancing its chances of survival (Degenhardt et al., 2003; Kappers et al., 2005). Beyond direct chemical defences, plants have devised an ingenious method of indirect defence through the release of VOCs (Heil, 2008). Damaged plants emit a complex blend of VOCs into the air, serving as a distress call to natural enemies of herbivores, such as predators and parasitoids. This airborne communication network not only attracts herbivore predators but also influences neighbouring plants, prompting them to activate their own defences as a pre-emptive measure against potential herbivore attacks (Sobhy et al., 2014). Understanding these inducible defences contributes not only to our comprehension of plant survival tactics but also holds potential implications for sustainable pest management and ecosystem balance.

4.7 PLANT DEFENCE SIGNALLING

When herbivores attack plants, they often induce a cascade of chemical responses, including the release of VOCs into the surrounding air. These volatile compounds, collectively termed herbivore-induced plant volatiles (HIPVs), act as crucial signalling molecules that mediate inter- and intra-plant communication (Holopainen and Blande, 2013). Neighbouring plants can detect these airborne cues, initiating defence responses even before physical contact with herbivores occurs. The perception of HIPVs involves specialised chemoreceptors on the surface of plant cells, which recognise and interpret specific volatile compounds. These receptors, often localised on plant leaves and stems, play a central role in initiating defence cascades. The presence of certain HIPVs can alert neighbouring plants to the impending threat of herbivory, prompting them to prepare their own chemical defences (Meents and Mithöfer, 2020).

The plant's response to herbivore attack is orchestrated through complex signalling networks. Among the key players are two well-studied hormone signalling pathways: jasmonic acid (JA) and salicylic acid (SA) (War et al., 2012, 2020). These pathways serve as central hubs in transmitting the signals initiated by the perception of HIPVs and other herbivore-induced cues. The JA pathway primarily governs defence responses against chewing herbivores and necrotrophic pathogens. It triggers the synthesis of defensive compounds such as protease inhibitors, secondary metabolites and PIs, all of which act as formidable deterrents to herbivores. The activation of the JA pathway often comes at the cost of growth and development, reflecting the plant's strategic allocation of resources toward defence (Onkokesung, Baldwin and Gális, 2010; Schweiger et al., 2014). Conversely, the SA pathway is associated with resistance against biotrophic pathogens and is involved in systemic acquired resistance (SAR). SA signalling prompts the accumulation of pathogenesis-related (PR) proteins, antimicrobial compounds and reactive oxygen species. Notably, the SA pathway can antagonise the JA pathway, leading to a trade-off between defence against pathogens and herbivores (Schweiger et al., 2014; Saleem, Fariduddin and Castroverde, 2021; Yang, Dolatabadian and Fernando, 2022).

The intricate relationship between the JA and SA pathways adds another layer of complexity to plant defence responses (Robert-Seilaniantz, Grant and Jones, 2011; Aerts, Pereira Mendes and Van Wees, 2021). Crosstalk between these pathways can either enhance or suppress certain defence mechanisms, depending on the specific ecological context and the nature of the threat. While the antagonism between JA and SA pathways is well-documented, recent research has unveiled instances of coordinated activation to maximise overall plant fitness (Aerts, Pereira Mendes and Van Wees, 2021). The regulation of this crosstalk involves a delicate balance between the expression of various transcription factors, hormone receptors and downstream defence genes. Environmental cues, such as herbivore species and intensity, dictate the priority of defence responses, leading to context-dependent activation of the JA and SA pathways' responses

(Robert-Seilaniantz, Grant and Jones, 2011; Aerts, Pereira Mendes and Van Wees, 2021). In conclusion, the mechanisms of perception and signalling in response to herbivore-induced cues constitute a sophisticated network that enables plants to sense impending threats and mount rapid defence responses. The dynamic interplay between the JA and SA pathways showcases the intricacies of plant defence strategies, in which the optimisation of resource allocation between growth and defence is finely tuned. Understanding these mechanisms sheds light on the remarkable adaptability of plants and provides a basis for developing targeted strategies in pest management and crop protection.

4.8 PLANT DEFENCE AND GROWTH TRADE-OFFS

The antagonism between the production of chemical defences and growth is posited as a natural consequence of adapting to terrestrial habitats during evolution. The dichotomy between growth and chemical defence is traditionally attributed to a limited pool of resources, suggesting that allocating resources to defence could impede growth and vice versa (Guo, Major and Howe, 2018; Sestari and Campos, 2022). Yet, this resource availability paradigm is increasingly viewed as overly simplistic. Recent molecular insights shed light on the intricate mechanisms that plants employ to fine-tune their phenotypes in response to diverse environmental challenges.

Metabolic costs associated with chemical defence production are substantial. Plants must allocate resources not only to the biosynthesis of defensive compounds but also to a complex biological machinery governing cellular processes such as modification, transport and storage (Züst and Agrawal, 2017; Erb and Kliebenstein, 2020). The resource availability paradigm is evolving into a more comprehensive theory that considers the dynamic interplay of genetic and metabolic networks. Plants demonstrate the ability to reconfigure their metabolic pathways to optimise growth versus defence conflicts in response to environmental signals (Bekaert et al., 2012). Crucially, hormonal transcriptional cascades, particularly those involving jasmonates (JAs) and salicylates (SAs), play a pivotal role in modulating resource allocation decisions. The crosstalk between these pathways provides a mechanistic explanation for the trade-off between chemical defence synthesis and growth. This hormonal interplay evolved as a response to the challenges posed by terrestrialisation, further contributing to the complexity of plant immune systems (Züst and Agrawal, 2017).

Another theory suggests that some plant chemical defences, while toxic to attacking organisms, can also harm the producing plant. Autotoxicity, a phenomenon in which a plant's own chemical defences negatively impact its growth and development, adds another layer to the growth–defence dilemma (Herms and Mattson, 1992; Neilson et al., 2013; Sestari and Campos, 2022). Plants have evolved self-resistance mechanisms to counteract autotoxicity, involving the biosynthesis or translocation of chemical defences, modification of toxic compounds and the formation of protective clusters (Baldwin and Callahan, 1993). Understanding the costs associated with chemical defence production has paved the way for innovative strategies to mitigate developmental consequences. Genetic rewiring of phytohormone transcriptional modules in model plants such as *Arabidopsis thaliana* offers a promising avenue. Mutations in JAZ genes, coupled with alterations in photoreceptor activity, showcase the potential to uncouple growth–defence trade-offs (Campos et al., 2016; Major et al., 2020).

Plants also form beneficial associations with microorganisms, such as the symbiosis between grasses and Epichloë fungal endophytes (Zamioudis and Pieterse, 2012; Finkel et al., 2017; Card, Bastías and Caradus, 2021; Marmolejo, Thompson and Helms, 2021). These associations not only promote growth but also contribute to plant defence by producing bioactive compounds. Additionally, root-associated microorganisms have been explored as tools to modulate growth and defences in response to environmental signals (Mitter et al., 2013). Defence priming, a physiological state induced by prior stress exposure, enhances the activation of defence responses against subsequent challenges. This strategy provides plants with a "state of readiness," resulting in a more robust and resource-effective response to future attacks (Conrath et al., 2015). Optimising metabolic routes associated with the synthesis of defensive barriers presents an elegant approach. Some plants produce multifunctional secondary metabolites, such as cyanogenic glucosides, serving dual purposes in both growth and defence processes (Erb and Kliebenstein, 2020). In conclusion, the cost of production of chemical defences in plants involves a complex interplay of genetic, metabolic and environmental factors. Unravelling these trade-offs has led to the identification of strategies that may empower plants to defend themselves without compromising their growth and overall fitness in dynamic ecosystems.

4.9 IMPACT OF GLOBAL CLIMATE CHANGE ON PLANT CHEMICAL DEFENCES

As we discussed previously in this chapter, the intricate mechanisms of plant chemical defence emerge as essential components in safeguarding plant health and resilience. However, it becomes evident that the dynamic relationship between plants and their herbivorous adversaries is significantly influenced by global climate changes (Bidart-Bouzat and Imeh-Nathaniel, 2008; DeLucia et al., 2012). Understanding the effect of climate changes on plant defence allows us to understand how external factors and environmental shifts play a pivotal role in shaping the intricate tapestry of plant–herbivore interactions, providing valuable insights into the adaptability and resilience of plant chemical defence strategies (DeLucia et al., 2012).

One significant aspect of global change is the rising atmospheric levels of carbon dioxide (CO_2), a consequence of human-induced activities. This surge has far-reaching implications for plant growth and ecosystems, potentially impacting plant–insect interactions. However, the effects are nuanced and contingent upon factors such as insect herbivory, plant competition, temperature, light, water stress and nutrient availability. Elevated CO_2 levels can induce alterations in plant traits, including changes in secondary metabolites crucial for plant defence (Robinson, Ryan and Newman, 2012). The intricate dynamics involve the carbon-nutrient balance hypothesis, explaining shifts in secondary metabolite concentrations and potential impacts on the inducibility of plant chemical defences, influencing plant–insect relationships. Another facet of global change is the increase in ozone (O_3) levels in the lower atmosphere, attributed to human activities. In contrast to CO_2, O_3 induces oxidative stress in plants, affecting various physiological processes (Himanen et al., 2008; Lindroth, 2010). Elevated O_3 levels influence plant secondary chemicals, with varied responses in terms of induced metabolites, subsequently affecting insect herbivores and their associated predators or parasitoids. The intricate interplay of elevated O_3 on plant secondary metabolism and its implications for disease resistance requires further investigation (Bidart-Bouzat and Imeh-Nathaniel, 2008; Himanen et al., 2008).

Concerns have also arisen regarding the consequences of stratospheric ozone depletion and increased ultraviolet (UV) radiation on terrestrial ecosystems in recent decades. UV light, essential for various plant processes, can also be harmful, causing damage to proteins, DNA and other biopolymers (Kunz et al., 2006). Higher UV levels induce an increase in phenolic compounds, especially flavonoids, influencing plant responses to stress. UV-B radiation, crucial for plant growth, also plays a role in plant defence responses to insect herbivory. Higher UV-B levels result in changes in plant secondary chemicals, impacting herbivory and insect performance. However, specificity in response to UV-B radiation necessitates further investigation into its impact on alkaloids and terpenoids, as well as its role in enhancing plant resistance to pathogens (Espinosa-Leal et al., 2022). Furthermore, elevated temperatures can also modify plant secondary chemistry, affecting glucosinolates, phenolic compounds and VOCs, with subsequent implications for plant–insect interactions.

4.10 CONCLUSION

In conclusion, the central role of plant chemical defence responses in mediating complex plant–insect interactions cannot be overstated. This intricate interplay influences not only the survival and behaviour of herbivores but also shapes entire ecosystems. Through an array of chemical compounds and signalling pathways, plants have evolved to defend against herbivores, thereby influencing the dynamics of trophic interactions and community structures. The exploration of these defence mechanisms has not only deepened our understanding of ecological relationships but also unlocked opportunities for practical applications. Continued research in this field is paramount for unravelling the intricacies of plant defences, offering insights that can revolutionise agriculture through sustainable pest management strategies. Moreover, as we confront global challenges of conservation and environmental change, a robust comprehension of plant chemical defence responses holds the potential to drive effective conservation initiatives and enhance our broader ecological understanding. The journey into the realm of plant–insect chemical interactions is an ongoing and vital pursuit with far-reaching implications for the betterment of both human society and the natural world.

REFERENCES

Abdalla, M. A. and Mühling, K. H. (2019) 'Plant-derived sulfur containing natural products produced as a response to biotic and abiotic stresses: a review of their structural diversity and medicinal importance', *Journal of Applied Botany and Food Quality*, 92, pp. 204–215.

Aerts, N., Pereira Mendes, M. and Van Wees, S. C. M. (2021) 'Multiple levels of crosstalk in hormone networks regulating plant defense', *The Plant Journal*, 105(2), pp. 489–504.

Agrawal, A. A. (2011) 'Current trends in the evolutionary ecology of plant defence', *FunctionalEcology*, 25(2), pp. 420–432. doi: 10.1111/j.1365-2435.2010.01796.x.

Agrawal, A. A. and Kurashige, N. S. (2003) 'A role for isothiocyanates in plant resistance against the specialist herbivore Pieris rapae', *Journal of Chemical Ecology*, 29, pp. 1403–1415.

Ahuja, I., Rohloff, J. and Bones, A. M. (2010) 'Defence mechanisms of Brassicaceae: implications for plant-insect interactions and potential for integrated pest management. A review', *Agronomy for Sustainable Development*, 30(2), pp. 311–348. doi: 10.1051/agro/2009025.

Ali, J. et al. (2021) 'Effects of cis-Jasmone treatment of brassicas on interactions with Myzus persicae Aphids and their parasitoid diaeretiella rapae', *Frontiers in Plant Science*, 12. doi: 10.3389/fpls.2021.711896.

Ali, J. et al. (2023) 'Exogenous application of Methyl Salicylate induces defence in brassica against peach potato Aphid Myzus persicae', *Plants*, 12(9), p. 1770.

Ali, J., Mukarram, M., et al. (2024) 'Wound to survive: mechanical damage suppresses aphid performance on brassica', *Journal of Plant Diseases and Protection*, 131, pp. 1–12.

Ali, J., Tonğa, A., et al. (2024) 'Defense strategies and associated phytohormonal regulation in brassica plants in response to chewing and Sap-sucking insects', *Frontiers in Plant Science*, 15, p. 1376917.

Ali, J. G. and Agrawal, A. A. (2012) 'Specialist versus generalist insect herbivores and plant defense', *Trends in Plant Science*, 17(5), pp. 293–302.

Aljbory, Z. and Chen, M. (2018) 'Indirect plant defense against insect herbivores: a review', *Insect Science*, 25(1), pp. 2–23.

Baldwin, I. T. and Callahan, P. (1993) 'Autotoxicity and chemical defense: nicotine accumulation and carbon gain in solanaceous plants', *Oecologia*, 94, pp. 534–541.

Bekaert, M. et al. (2012) 'Metabolic and evolutionary costs of herbivory defense: systems biology of glucosinolate synthesis', *New Phytologist*, 196(2), pp. 596–605.

Bennett, R. N. and Wallsgrove, R. M. (1994) 'Secondary metabolites in plant defence mechanisms', *New Phytologist*, 127(4), pp. 617–633.

Beran, F. and Petschenka, G. (2022) 'Sequestration of plant defense compounds by insects: from mechanisms to insect–plant coevolution', *Annual Review of Entomology*, 67, pp. 163–180.

Berenbaum, M. R. and Zangerl, A. R. (1998) 'Chemical phenotype matching between a plant and its insect herbivore', *Proceedings of the National Academy of Sciences*, 95(23), pp. 13743–13748.

Bidart-Bouzat, M. G. and Imeh-Nathaniel, A. (2008) 'Global change effects on plant chemical defenses against insect herbivores', *Journal of Integrative Plant Biology*, 50(11), pp. 1339–1354. doi: 10.1111/j.1744-7909.2008.00751.x.

Birnbaum, S. S. L. and Abbot, P. (2018) 'Insect adaptations toward plant toxins in milkweed–herbivores systems–a review', *Entomologia Experimentalis et Applicata*, 166(5), pp. 357–366.

Boots, M. and Best, A. (2018) 'The evolution of constitutive and induced defences to infectious disease', *Proceedings of the Royal Society B: Biological Sciences*, 285(1883), p. 20180658.

Bruce, T. J. A. (2014) 'Glucosinolates in oilseed rape: secondary metabolites that influence interactions with herbivores and their natural enemies', *Annals of Applied Biology*, 164(3), pp. 348–353.

Bruce, T. J. A. (2015) 'Interplay between insects and plants: dynamic and complex interactions that have coevolved over millions of years but act in milliseconds', *Journal of Experimental Botany*, 66(2), pp. 455–465. doi: 10.1093/jxb/eru391.

Bui, V.-H., Rodríguez-López, C. E. and Dang, T.-T. T. (2023) 'Integration of discovery and engineering in plant alkaloid research: recent developments in elucidation, reconstruction, and repurposing biosynthetic pathways', *Current Opinion in Plant Biology*, 74, p. 102379.

Campos, M. L. *et al.* (2016) 'Rewiring of jasmonate and phytochrome B signalling uncouples plant growth-defense tradeoffs', *Nature Communications*, 7(1), p. 12570.

Card, S. D., Bastías, D. A. and Caradus, J. R. (2021) 'Antagonism to plant pathogens by Epichloë fungal endophytes—A review', *Plants*, 10(10), p. 1997.

Chen, M. S. (2008) 'Inducible direct plant defense against insect herbivores: a review', *Insect Science*, 15(2), pp. 101–114. doi: 10.1111/j.1744-7917.2008.00190.x.

Chhajed, S. *et al.* (2019) 'Chemodiversity of the glucosinolate-myrosinase system at the single cell type resolution', *Frontiers in Plant Science*, 10, p. 618.

Conrath, U. *et al.* (2015) 'Priming for enhanced defense', *Annual Review of Phytopathology*, 53(1), pp. 97–119. doi: 10.1146/annurev-phyto-080614-120132.

Croteau, R., Kutchan, T. M. and Lewis, N. G. (2000) 'Natural products (secondary metabolites)', *Biochemistry and Molecular Biology of Plants*, 24, pp. 1250–1319.

Crozier, A., Clifford, M. N. and Ashihara, H. (2006) 'Plant secondary metabolites', *Occurrence, structure and role in the human diet*. Blackwell-Publishers.

Crozier, A., Jaganath, I. B. and Clifford, M. N. (2006) 'Phenols, polyphenols and tannins: an overview', *Plant secondary metabolites: occurrence, structure and role in the human diet*, 1, pp. 1–25.

Dambolena, J. S. *et al.* (2016) 'Terpenes: natural products for controlling insects of importance to human health—A structure-activity relationship study', *Psyche: A Journal of Entomology*, 2016, pp. 1–17.

Degenhardt, J. *et al.* (2003) 'Attracting friends to feast on foes: engineering terpene emission to make crop plants more attractive to herbivore enemies', *Current Opinion in Biotechnology*, 14(2), pp. 169–176. doi: 10.1016/S0958-1669(03)00025-9.

Degenhardt, J., Köllner, T. G. and Gershenzon, J. (2009) 'Monoterpene and ses-quiterpene synthases and the origin of terpene skeletal diversity in plants', *Phytochemistry*, 70(15–16), pp. 1621–1637.

DeLucia, E. H. *et al.* (2012) 'Climate change: resetting plant-insect interactions', *Plant Physiology*, 160(4), pp. 1677–1685.

Dicke, M. and Hilker, M. (2003) 'Induced plant defences: from molecular biology to evolutionary ecology', *Basic and Applied Ecology*, 4(1), pp. 3–14.

Divekar, P. A. *et al.* (2022) 'Plant secondary metabolites as defense tools against herbivores for sustainable crop protection', *International Journal of Molecular Sciences*, 23(5), p. 2690.

Divekar, P. A. *et al.* (2023) 'Protease inhibitors: an induced plant defense mecha-nism against herbivores', *Journal of Plant Growth Regulation*, 42(10), pp. 6057–6073.

Divekar, P., Kumar, P. and Suby, S. B. (2019) 'Screening of maize germplasm through antiobiosis mechanism of resistance against Chilo partellus (Swinhoe)', *Journal of Entomology and Zoology Studies*, 7, pp. 1111–1114.

Erb, M. and Kliebenstein, D. J. (2020) 'Plant secondary metabolites as defenses, regulators, and primary metabolites: the blurred functional trichotomy1[OPEN]', *Plant Physiology*, 184(1), pp. 39–52. doi: 10.1104/PP.20.00433.

Espinosa-Leal, C. A. *et al.* (2022) 'Recent advances on the use of abiotic stress (water, UV radiation, atmospheric gases, and temperature stress) for the enhanced production of secondary metabolites on in vitro plant tissue cul-ture', *Plant Growth Regulation*, 97(1), pp. 1–20.

Falcone Ferreyra, M. L., Rius, S. P. and Casati, P. (2012) 'Flavonoids: biosynthesis, biological functions, and biotechnological applications', *Frontiers in Plant Science*, 3, p. 222.

Finkel, O. M. *et al.* (2017) 'Understanding and exploiting plant beneficial microbes', *Current Opinion in Plant Biology*, 38, pp. 155–163.

Frei, M. (2013) 'Lignin: characterization of a multifaceted crop component', *The Scientific World Journal*, 2013(1), pp. 436–517.

Fürstenberg-Hägg, J., Zagrobelny, M. and Bak, S. (2013) 'Plant defense against insect herbivores', *International Journal of Molecular Sciences*, 14(5), pp. 10242–10297.

Gershenzon, J. and Dudareva, N. (2007) 'The function of terpene natural prod-ucts in the natural world', *Nature Chemical Biology*, 3(7), pp. 408–414.

Guo, Q., Major, I. T. and Howe, G. A. (2018) 'Resolution of growth–defense con-flict: mechanistic insights from jasmonate signaling', *Current Opinion in Plant Biology*, 44, pp. 72–81.

Hadacek, F. (2002) 'Secondary metabolites as plant traits: current assessment and future perspectives', *Critical Reviews in Plant Sciences*, 21(4), pp. 273–322.

Halkier, B. A. and Gershenzon, J. (2006) 'Biology and biochemistry of glucosino-lates', *Annual Review of Plant Biology*, 57, pp. 303–333.

Haslam, E. (2014) *The shikimate pathway: biosynthesis of natural products series.* Elsevier.

Hassanpour, S., MaheriSis, N., Eshratkhah, B. and Baghbani Mehmandar, F. (2011) 'Plants and secondary metabolites (Tannins): a review'. *International Journal of Forest, Soil and Erosion*, 1, pp. 47–53.

Heil, M. (2008) 'Indirect defence via tritrophic interactions', *New Phytologist*, 178(1), pp. 41–61. doi: 10.1111/j.1469-8137.2007.02330.x.

Herms, D. A. and Mattson, W. J. (1992) 'The dilemma of plants: to grow or defend', *The Quarterly Review of Biology*, 67(3), pp. 283–335.

Himanen, S. J. *et al.* (2008) 'Constitutive and herbivore-inducible glucosinolate concentrations in oilseed rape (*Brassica napus*) leaves are not affected by Bt Cry1Ac insertion but change under elevated atmospheric CO_2 and O_3', *Planta*, 227, pp. 427–437.

Holopainen, J. K. and Blande, J. D. (2013) 'Where do herbivore-induced plant volatiles go?', *Frontiers in Plant Science*, 4, p. 185.

Hopkins, R. J., van Dam, N. M. and van Loon, J. J. A. (2009) 'Role of glucosinolates in insect-plant relationships and multitrophic interactions', *Annual Review of Entomology*, 54, pp. 57–83.

Isman, M. B. and Paluch, G. (2011) 'Needles in the haystack: exploring chemical diversity of botanical insecticides', *Green trends in insect control*. Royal Society of Chemistry, pp. 248–265.

James, D. G. *et al.* (2012) 'Employing chemical ecology to understand and exploit biodiversity for pest management', *Biodiversity and insect pests: key issues for sustainable management*, pp. 185–195. doi: 10.1002/9781118231838.ch11.

Jander, G. and Howe, G. (2008) 'Plant interactions with arthropod herbivores: state of the field', *Plant Physiology*, 146(3), pp. 801–803. doi: 10.1104/pp.104.900247.

Jones, P. and Vogt, T. (2001) 'Glycosyltransferases in secondary plant metabolism: tranquilizers and stimulant controllers', *Planta*, 213, pp. 164–174.

Jung, H. G. and Allen, M. S. (1995) 'Characteristics of plant cell walls affecting intake and digestibility of forages by ruminants', *Journal of Animal Science*, 73(9), pp. 2774–2790.

Kappers, I. F. *et al.* (2005) 'Genetic engineering of terpenoid metabolism attracts bodyguards to Arabidopsis', *Science*, 309(5743), pp. 2070–2072. doi: 10.1126/science.1116232.

Kitainda, V. and Jez, J. M. (2021) 'Structural studies of aliphatic glucosinolate chain-elongation enzymes', *Antioxidants*, 10(9), p. 1500.

Koshiyama, M. *et al.* (2006) 'Development of a new plant growth regulator, prohydrojasmon', *Regulation of Plant Growth and Development*, 41, pp. 24–33.

Kumar, J. *et al.* (2023) 'A comprehensive review on examination of plant biochemistry including pathways, enzymes, and applications', *International Journal of Plant & Soil Science*, 35(21), pp. 1192–1201.

Kumar, P. and Suby, S. B. (2013) 'Antibiosis effect of phenolic acids (ferulic acid and P-Coumaric acid) on maize spotted stem borer, Chilo partellus (Swinhoe)(lepidoptera: pyralidae)', *Indian Journal of Entomology*, 75(3), pp. 247–250.

Kunz, B. A. *et al.* (2006) 'Plant responses to UV radiation and links to pathogen resistance', *International Review of Cytology*, 255, pp. 1–40.

Lattanzio, V. *et al.* (2009) 'Plant phenolics—secondary metabolites with diverse functions', *Recent Advances in Polyphenol Research*, 1, pp. 1–35.

Lattanzio, V. (2013) 'Phenolic compounds: introduction 50', *Natural products*, pp. 1543–1580.

Lichman, B. R. (2021) 'The scaffold-forming steps of plant alkaloid biosynthesis', *Natural Product Reports*, 38(1), pp. 103–129.

Lin, D. *et al.* (2016) 'An overview of plant phenolic compounds and their importance in human nutrition and management of type 2 diabetes', *Molecules*, 21(10), p. 1374.

Lindroth, R. L. (2010) 'Impacts of elevated atmospheric CO_2 and O_3 on forests: phytochemistry, trophic interactions, and ecosystem dynamics', *Journal of Chemical Ecology*, 36, pp. 2–21.

Lv, Q. *et al.* (2022) 'The cellular and subcellular organization of the glucosinolate–myrosinase system against herbivores and pathogens', *International Journal of Molecular Sciences*, 23(3), p. 1577.

Macel, M. (2011) 'Attract and deter: a dual role for pyrrolizidine alkaloids in plant–insect interactions', *Phytochemistry Reviews*, 10, pp. 75–82.

Major, I. T. *et al.* (2020) 'A phytochrome B-independent pathway restricts growth at high levels of jasmonate defense', *Plant Physiology*, 183(2), pp. 733–749.

Al Mamari, H. H. (2021) 'Phenolic compounds: classification, chemistry, and updated techniques of analysis and synthesis', *Phenolic compounds: chemistry, synthesis, diversity, non-conventional industrial, pharmaceutical and therapeutic applications*, pp. 73–94.

Marmolejo, L. O., Thompson, M. N. and Helms, A. M. (2021) 'Defense suppression through interplant communication depends on the attacking herbivore species', *Journal of Chemical Ecology*, 47(12), pp. 1049–1061.

Mathur, A. *et al.* (2024) 'Regulating pri/pre-microRNA up/down expressed in cancer proliferation, angiogenesis and metastasis using selected potent triterpenoids', *International Journal of Biological Macromolecules*, 257, p. 127945.

Matsuura, H. N. and Fett-Neto, A. G. (2015) 'Plant alkaloids: main features, toxicity, and mechanisms of action', *Plant Toxins*, 2(7), pp. 1–15.

McKey, D. (1979) 'The distribution of secondary compounds within plants', *Herbivores-their interaction with secondary plant metabolites*, pp. 55–134.

Meents, A. K. and Mithöfer, A. (2020) 'Plant–plant communication: is there a role for volatile damage-associated molecular patterns?', *Frontiers in Plant Science*, 11, p. 583275.

Mierziak, J., Kostyn, K. and Kulma, A. (2014) 'Flavonoids as important molecules of plant interactions with the environment', *Molecules*, 19(10), pp. 16240–16265.

Mithen, R. (2008) 'Sulphur-containing compounds', *Plant secondary metabolites: occurrence, structure and role in the human diet*, pp. 25–46.

Mithöfer, A. and Boland, W. (2012) 'Plant defense against herbivores: chemical aspects', *Annual Review of Plant Biology*, 63, pp. 431–450. doi: 10.1146/annurev-arplant-042110-103854.

Mithöfer, A., Boland, W. and Maffei, M. E. (2009) 'Chemical ecology of plant-insect interactions', *Molecular aspects of plant disease resistance.* Wiley-Blackwell, pp. 261–291.

Mitter, B. *et al.* (2013) 'Advances in elucidating beneficial interactions between plants, soil, and bacteria', *Advances in Agronomy*, 121, pp. 381–445.

Moore, K. J. and Jung, H.-J. G. (2001) 'Lignin and fiber digestion'. *Journal of Range Management*, 54(4), pp. 420–430.

Mumm, R., Posthumus, M. A. and Dicke, M. (2008) 'Significance of terpenoids in induced indirect plant defence against herbivorous arthropods', *Plant, Cell & Environment*, 31(4), pp. 575–585.

Neilson, E. H. *et al.* (2013) 'Plant chemical defense: at what cost?', *Trends in Plant Science*, 18(5), pp. 250–258.

Ninkuu, V. *et al.* (2021) 'Biochemistry of terpenes and recent advances in plant protection', *International Journal of Molecular Sciences*, 22(11), p. 5710.

Onkokesung, N., Baldwin, I. T. and Gális, I. (2010) 'The role of jasmonic acid and ethylene crosstalk in direct defense of Nicotiana attenuata plants against chewing herbivores', *Plant Signaling & Behavior*, 5(10), pp. 1305–1307.

Paré, P. W. and Tumlinson, J. H. (1999) 'Plant volatiles as a defense against insect herbivores', *Plant Physiology*, 121(2), pp. 325–331. doi: 10.1104/pp.121.2.325.

Pentzold, S. *et al.* (2015) 'Metabolism, excretion and avoidance of cyanogenic glucosides in insects with different feeding specialisations', *Insect Biochemistry and Molecular Biology*, 66, pp. 119–128.

Pichersky, E. and Raguso, R. A. (2018) 'Why do plants produce so many terpenoid compounds?', *New Phytologist*, 220(3), pp. 692–702.

Poulton, J. E. (1983) 'Cyanogenic compounds in plants and their toxic effects', *Handbook of Natural Toxins*, 1, pp. 117–157.

Pratyusha, S. (2022) 'Phenolic compounds in the plant development and defense: an overview', *Plant stress physiology-perspectives in agriculture*, pp. 125–140.

Von Rad, U. *et al.* (2001) 'Two glucosyltransferases are involved in detoxification of benzoxazinoids in maize', *The Plant Journal*, 28(6), pp. 633–642.

Ramos, M. V. *et al.* (2019) 'Laticifers, latex, and their role in plant defense', *Trends in Plant Science*, 24(6), pp. 553–567.

Ratzka, A. *et al.* (2002) 'Disarming the mustard oil bomb', *Proceedings of the National Academy of Sciences*, 99(17), pp. 11223–11228.

Rehman, F., Khan, F. A. and Badruddin, S. M. A. (2012) 'Role of phenolics in plant defense against insect herbivory', *Chemistry of phytopotentials: health, energy and environmental perspectives*, pp. 309–313.

Renwick, J. A. A. *et al.* (2006) 'Isothiocyanates stimulating oviposition by the diamondback moth, Plutella xylostella', *Journal of Chemical Ecology*, 32, pp. 755–766.

Robert-Seilaniantz, A., Grant, M. and Jones, J. D. G. (2011) 'Hormone crosstalk in plant disease and defense: more than just jasmonate-salicylate antagonism', *Annual Review of Phytopathology*, 49, pp. 317–343.

Robinson, E. A., Ryan, G. D. and Newman, J. A. (2012) 'A meta-analytical review of the effects of elevated CO2 on plant–arthropod interactions highlights the importance of interacting environmental and biological variables', *New Phytologist*, 194(2), pp. 321–336.

Saleem, M., Fariduddin, Q. and Castroverde, C. D. M. (2021) 'Salicylic acid: a key regulator of redox signalling and plant immunity', *Plant Physiology and Biochemistry*, 168, pp. 381–397.

Samanta, A., Das, G. and Das, S. K. (2011) 'Roles of flavonoids in plants', *Carbon*, 100(6), pp. 12–35.

Santos-Sánchez, N. F. *et al.* (2019) 'Shikimic acid pathway in biosynthesis of phenolic compounds', *Plant physiological aspects of phenolic compounds*, 1, pp. 1–15.

Schultz, J. C. (1989) 'Tannin-insect interactions', *Chemistry and significance of condensed tannins*. Springer, pp. 417–433.

Schweiger, R. *et al.* (2014) 'Interactions between the jasmonic and salicylic acid pathway modulate the plant metabolome and affect herbivores of different feeding types', *Plant, Cell & Environment*, 37(7), pp. 1574–1585. doi: 10.1111/pce.12257.

Scott, J. G. and Wen, Z. (2001) 'Cytochromes P450 of insects: the tip of the iceberg', *Pest Management Science*, 57(10), pp. 958–967.

Selmar, D. (2010) 'Biosynthesis of cyanogenic glycosides, glucosinolates and nonprotein amino acids', *Annual plant reviews volume 40: biochemistry of plant secondary metabolism*, pp. 92–181.

Sestari, I. and Campos, M. L. (2022) 'Into a dilemma of plants: the antagonism between chemical defenses and growth', *Plant Molecular Biology*, 109(4–5), pp. 469–482.

Simmonds, M. S. J. (2003) 'Flavonoid–insect interactions: recent advances in our knowledge', *Phytochemistry*, 64(1), pp. 21–30.

Singh, A. (2017) 'Glucosinolates and plant defense', *Glucosinolates*. Springer International Publishing, pp. 237–246.

Singh, S. K. (2018) 'Explorations of plant's chemodiversity: role of nitrogen-containing secondary metabolites in plant defense', *Molecular aspects of plant-pathogen interaction*, pp. 309–332.

Sobhy, I. S. *et al.* (2014) 'The prospect of applying chemical elicitors and plant strengtheners to enhance the biological control of crop pests', *Philosophical Transactions of the Royal Society B: Biological Sciences*, 369(1639), p. 20120283. doi: 10.1098/rstb.2012.0283.

Srivasatava, P. (2022) 'Use of alkaloids in plant protection', *Plant protection: from chemicals to biologicals*, p. 337.

Tamaoki, D. *et al.* (2013) 'Jasmonic acid and salicylic acid activate a common defense system in rice', *Plant Signaling & Behavior*, 8(6), p. e24260.

Toffolatti, S. L. *et al.* (2021) 'Role of terpenes in plant defense to biotic stress', *Biocontrol agents and secondary metabolites*. Elsevier, pp. 401–417.

Uddin, N. *et al.* (2024) 'Lignin developmental patterns and Casparian strip as apoplastic barriers: a review', *International Journal of Biological Macromolecules*, 260, p. 129595.

Vallad, G. E. and Goodman, R. M. (2004) 'Systemic acquired resistance and induced systemic resistance in conventional agriculture', *Crop Science*, 44, pp. 1920–1934.

Vucetic, A. *et al.* (2014) 'Volatile interaction between undamaged plants affects tritrophic interactions through changed plant volatile emission', *Plant Signaling & Behavior*, 9(8), p. e29517.

War, A. R. *et al.* (2012) 'Mechanisms of plant defense against insect herbivores', *Plant Signaling and Behavior*, 7(10). doi: 10.4161/psb.21663.

War, A. R. *et al.* (2018) 'Plant defence against herbivory and insect adaptations', *AoB Plants*, 10(4), p. ply037.

War, A. R. *et al.* (2020) 'Plant defense and insect adaptation with reference to secondary metabolites', *Co-evolution of secondary metabolites*, pp. 795–822.

Wetzel, W. C. *et al.* (2023) 'Variability in plant–herbivore interactions', *Annual Review of Ecology, Evolution, and Systematics*, 54, pp. 451–474.

Wink, M. (2017) 'The role of quinolizidine alkaloids in plant-insect interactions', *Insect-plant interactions (1992)*. CRC Press, pp. 139–174.

Wink, M., Schmeller, T. and Latz-Brüning, B. (1998) 'Modes of action of allelo-chemical alkaloids: interaction with neuroreceptors, DNA, and other molecular targets', *Journal of Chemical Ecology*, 24, pp. 1881–1937.

Wittstock, U. and Gershenzon, J. (2002) 'Constitutive plant toxins and their role in defense against herbivores and pathogens', *Current Opinion in Plant Biology*, 5(4), pp. 300–307.

Wittstock, U. *et al.* (2004) 'Successful herbivore attack due to metabolic diver-sion of a plant chemical defense', *Proceedings of the National Academy of Sciences*, 101(14), pp. 4859–4864.

Yang, C., Dolatabadian, A. and Fernando, W. G. D. (2022) 'The wonderful world of intrinsic and intricate immunity responses in plants against pathogens', *Canadian Journal of Plant Pathology*, 44(1), pp. 1–20.

Yates, A. D. and Michel, A. (2018) 'Mechanisms of aphid adaptation to host plant resistance', *Current Opinion in Insect Science*, 26, pp. 41–49.

Zamioudis, C. and Pieterse, C. M. J. (2012) 'Modulation of host immunity by ben-eficial microbes', *Molecular Plant-Microbe Interactions*, 25(2), pp. 139–150.

Zangerl, A. R. and Rutledge, C. E. (1996) 'The probability of attack and patterns of constitutive and induced defense: a test of optimal defense theory', *The American Naturalist*, 147(4), pp. 599–608.

Züst, T. and Agrawal, A. A. (2017) 'Trade-offs between plant growth and defense against insect herbivory: an emerging mechanistic synthesis', *Annual Review of Plant Biology*, 68(1), pp. 513–534.

CHAPTER 5

Host Plant Selection by Insect Herbivores

5.1 INTRODUCTION

In insect–plant interactions, the host plant emerges as a central protagonist, wielding unparalleled influence over the survival, reproduction and ecological dynamics of insect populations. The significance of host plants in shaping these interactions stems from the exquisite specialisation and evolution observed in insects, which have finely tuned their biology to feed on specific plant species. This chapter seeks to unravel the multifaceted dimensions of host plant selection by insect herbivores. Host plants wield a profound impact on the delicate balance of insect life histories and feeding strategies (Mello and Silva-Filho, 2002). Insects, with their highly evolved adaptations, are discerning consumers, relying on the chemical composition of their chosen host plants to navigate taste, nutritional value and toxicity (Ali, 2022, 2023). The plant's arsenal of phytochemicals, nutrients and secondary metabolites plays a pivotal role in shaping the intricate dance between herbivores and their chosen sustenance. Some insects exhibit remarkable detoxification abilities and tolerance to these chemical compounds, while others remain vulnerable, a testament to the dynamic arms race between plants and herbivores (Awmack and Leather, 2002).

The connection between host plants and herbivorous insects transcends mere sustenance; it extends to vital life processes such as shelter and oviposition. The quality status of host plants emerges as a critical

DOI: 10.1201/9781003479857-5

factor influencing the fecundity and physiology of herbivorous insects. The reproductive success of herbivorous insects hinges on the delicate balance of host plant quality, in which factors such as optimal oviposi-tional preferences and larval performance are intricately linked to the host plant's attributes (Price, 1999). Furthermore, the chapter explores how the presence of specific host plants reverberates through ecosystems, influenc-ing the abundance and diversity of insect species. Insect–plant interac-tions extend beyond individual survival strategies, as they significantly impact plant growth, reproduction and overall community structure. Understanding the role of host plants becomes paramount in managing and comprehending insect pest populations, offering insights into the development of sustainable pest management strategies (Price, 1999).

As we explore the chemical ecology of these interactions, the volatile compound emissions emerge as vital cues guiding insect foraging behav-iour. The intricate interplay of volatiles shapes the ability of insects to locate their host plants, emphasising the intricate connections between chemical signals and the ecological success of herbivores (Bruce and Pickett, 2011; Webster and Cardé, 2017). This chapter explores the complex process of how herbivorous insects select their host plant. It unravels the threads connecting feeding preferences, shelter requirements and ovipositional strategies while shedding light on how the quality status of host plants orchestrates the physiological performance and fecundity of herbivorous insects. Through this exploration, we aim to deepen our understanding of the chemical ecology that governs the complex dance between insects and their chosen botanical partners.

5.2 FACTORS INFLUENCING HOST PLANT SELECTION

Factors influencing host plant selection are greatly shaped by chemical cues. These cues are pivotal in guiding insect behaviour, impacting criti-cal decisions such as feeding, mating and oviposition. Understanding the intricate chemical language of plants provides insights into insect–plant interactions, leading to innovative pest management strategies and improved agricultural practices (Figure 5.1)

5.2.1 Plant Chemistry

Plant chemistry plays a crucial role in the decision-making process of her-bivores when selecting a host plant. The intricate dynamics between herbi-vore behaviour and plant chemistry unveils the multifaceted nature of host plant selection in the challenging environments where these interactions

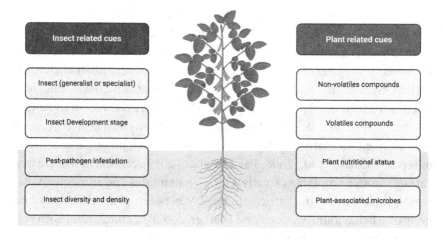

FIGURE 5.1   Factors influencing host plant selection: Insect-related and plant-related cues.

unfold (Wang *et al.*, 2020). In the context of host plant selection, chemical signals act as key mediators, influencing various facets of herbivore behaviour. Behavioural responses to chemical cues encompass a spectrum of actions, including attraction, orientation, feeding and oviposition, all finely tuned to respond to the chemical cues emitted by host plants. This chemical guidance aids herbivores in locating suitable feeding and reproductive sites within the complex array of plant species (Wang et al., 2020; Anton and Cortesero, 2022).

Insect herbivores possess specialised sensory mechanisms and olfactory systems, with olfactory receptors adept at detecting and distinguishing chemical compounds released by plants, even in trace amounts. This refined sensory machinery enables herbivores to discern intricate chemical profiles, differentiating between potential hosts and non-hosts (Rani, 2015; Conchou et al., 2019). Moreover, plant primary and secondary metabolisms play a pivotal role in determining host plant quality and can significantly impact the feeding performance and life history traits of phytophagous insects (Mumm, Posthumus and Dicke, 2008; Wang et al., 2020). Defensive chemicals such as alkaloids, glucosinolates, terpenoids and phenolics, known for their toxic, antifeedant or repellent properties, can markedly reduce insect performance (Axel Mithöfer and Boland, 2012; Al-Khayri et al., 2023). Conversely, higher concentrations of nitrogen in plants are beneficial for insect development (Chen, Olson and Ruberson, 2010). The structural components of plant cell walls, namely cellulose and lignin, act as constitutive barriers in plant defences (Kitajima et al., 2012; Yadav and Chattopadhyay, 2023). For instance, cellulose content effectively predicts plant defence traits, such as leaf toughness, influencing insect feeding patterns (Fuentealba et al., 2020). Specific phytochemicals, such as α-chaconine and caffeic acid in potato tuber skin, have been identified as influencing the survival of potato tuber worm larvae. In tomatoes, biochemical factors such as α-tomatine, as well as physical properties such as fruit size and maturity, impact the growth, development and survival of larvae (Mulatu et al., 2006; Pacifico et al., 2019). Similarly, leaf surface extracts from host plants can affect larval hatch rates and subsequent dispersal, highlighting the intricate interplay between plant chemistry and herbivore behaviour (Awmack and Leather, 2002). In conclusion, both volatile and non-volatile phytochemicals orchestrate a symphony of signals, influencing the delicate dance between herbivore behaviour and host plant selection in challenging environments.

In addition to volatiles, leaf surface compounds play a pivotal role in the intricate dynamics of host plant selection by herbivorous insects. These compounds, often referred to as epicuticular waxes or cuticular chemicals, constitute a diverse array of molecules that coat the outer layer of plant leaves (Anton and Cortesero, 2022). While initially thought to primarily serve as a barrier against water loss and pathogens, it has become evident that leaf surface compounds play a significant role in mediating interactions between plants and herbivores. These compounds can act as vital cues that herbivores use to identify and evaluate potential host plants for feeding and oviposition. For instance, in *Bombyx mori*, specialised coumarin receptor neurons that express the gustatory receptors BmGr53 and BmGr19 are located on the larval maxillary palp. These neurons exhibit high sensitivity and are capable of detecting even trace amounts of the toxic compound on the leaf surface of non-host plants, such as the cherry plant. In response to this detection, the larvae suppress test biting, effectively avoiding the ingestion of potentially harmful compounds (Shii et al., 2021).

The chemical composition of leaf surface compounds can vary greatly across plant species and even among individuals of the same species. This variation is attributed to a complex interplay of genetic factors, developmental stages and environmental conditions (Müller and Riederer, 2005). Herbivores have evolved a remarkable ability to detect and discriminate among these compounds, allowing them to make informed decisions about the suitability of a host plant. Some herbivores, especially specialists that have coevolved with specific plant lineages, can recognise even sub-tle variations in leaf surface chemicals (Fernández et al., 2019; Shii et al., 2021). For instance, specific trichomes, glandular structures that release chemical compounds, can act as beacons, emitting volatile compounds that signal the presence of a suitable host plant (Levin, 1973; Tissier, Morgan and Dudareva, 2017). In contrast, certain chemicals on the leaf surface may deter herbivores, either due to their toxicity or unpalatability (Levin, 1973; Dalin et al., 2008). These compounds provide a clear "do not feed" signal to herbivores and influence their host plant preferences. In summary, leaf surface compounds contribute to the complex puzzle of host plant selection by herbivores. Their chemical diversity, variability and role in communication underscore their significance in the ongoing interactions between plants and herbivores.

5.2.2 Role of Plant-Produced Volatiles

Plants release volatile compounds that function as chemical signals, conveying valuable information regarding the host plant's physiologi-cal condition, the presence of potential herbivores and the availability of resources (Pickett and Khan, 2016; Ninkovic, Markovic and Rensing, 2021). Volatiles are emitted from various plant tissues, including leaves, stems, flowers and fruits, and their composition can vary widely based on factors such as plant species, developmental stage and environmental con-ditions. Volatiles play a crucial role in attracting herbivores to their host plants over varying distances. Many herbivores possess highly sensitive olfactory systems that allow them to detect and respond to minute quanti-ties of volatiles released by plants (Anderson and Anton, 2014; Arimura, 2021). These volatiles act as olfactory beacons, guiding herbivores toward suitable hosts. For example, female moths of the tobacco hornworm, *Manduca sexta*, are attracted to the scent of specific tobacco plants that emit volatiles signalling their suitability as oviposition sites (Mechaber, Capaldo and Hildebrand, 2002; Späthe, 2014).

The chemical diversity of emitted volatiles contributes to the specificity of herbivore–plant interactions. Different plant species emit unique blends of volatiles, each with distinct chemical profiles. Herbivores have evolved to recognise and respond to these specific blends, enabling them to dif-ferentiate between potential host plants and non-hosts (McCormick et al., 2019; Anton and Cortesero, 2022). This chemical specificity enhances the efficiency of host plant recognition and reduces the likelihood of herbi-vores wasting energy on unsuitable plants. Plants can alter their volatile emission patterns in response to herbivore damage or other stressors. This phenomenon, known as induced volatile emission, is a dynamic response that enhances the plant's ability to communicate with herbivores and natural enemies (Blande, Holopainen and Niinemets, 2014; Faiola and Taipale, 2020). When a plant is attacked, it may release a different set of volatiles that attract predators or parasitoids of the herbivore, contributing to a complex web of tritrophic interactions (Turlings and Erb, 2018).

Conclusively, plant-produced volatiles play a central role in the intricate dance of host plant selection by insect herbivores. Their chemical diversity, specificity and ability to attract herbivores from a distance underscore their significance in mediating interactions between plants and herbivores (Turlings and Erb, 2018). By exploiting these volatile cues, herbivores optimise their foraging and reproductive strategies, while plants gain a means to communicate with the herbivore community and enlist the assistance of natural enemies. Understanding the intricate mechanisms underlying the role of volatiles in host plant selection enriches our comprehension of the chemical ecology that underpins these complex interactions (McCormick, Unsicker and Gershenzon, 2012).

5.2.3 Nutritional Quality and Defence-Related Cues

In the intricate process of host plant selection by herbivorous insects, nutritional quality and defence-related cues play pivotal roles. Herbivores are finely attuned to the chemical signals emitted by plants, which convey critical information about the plant's nutritional value and defensive capabilities (Ninkovic, Markovic and Rensing, 2021). Nutrient-rich plants often release specific volatile compounds that attract herbivores in search of optimal feeding sites. These cues are essential for herbivores to make informed decisions about their food sources, contributing to their growth and reproduction (Chen, Olson and Ruberson, 2010). Conversely, defensive chemicals, such as secondary metabolites, act as deterrents, signalling to herbivores the presence of potentially harmful compounds (A Mithöfer and Boland, 2012; Pagare et al., 2015). Herbivores must strike a balance between accessing nutrition and avoiding toxic or unpalatable substances, a process heavily influenced by the chemical cues emitted by plants (Karban and Baldwin, 2007; Parikh et al., 2017). As herbivores navigate the complex landscape of plant species, the interplay between nutritional quality and defence-related cues shapes their host plant selection strategies, with profound implications for their survival and ecological interactions (Burkepile and Parker, 2017; Wetzel et al., 2023).

5.2.4 Plant-Associated Microbes

Insects, acting as herbivores, navigate ecosystems in search of suitable host plants for their growth and reproduction, relying on plant-produced cues. The intricate interplay of plants, herbivores and microorganisms, both beneficial and pathogenic, adds complexity to herbivore foraging. This section delves into the roles of plant-associated microbes in shaping the visual, olfactory and gustatory cues that guide insect herbivores in their host plant selection process (Grunseich et al., 2019). Visual cues, encompassing patterns, dimensions and spectral quality, play a pivotal role in herbivorous insects foraging by conveying information about plant location, nutrition and defence status. Beneficial microbes, exemplified by arbuscular mycorrhizal fungi (AMF) and plant growth-promoting rhizobacteria (PGPR), influence visual cues through their impact on plant growth and biomass, potentially affecting herbivore foraging and oviposition behaviour (Grunseich et al., 2019; Raj et al., 2022). Additionally, pathogenic microbes alter visual cues by inducing disease symptoms that modify physical plant traits, influencing both vector and non-vector herbivores (Shikano et al., 2017). Understanding these microbial influences on visual cues provides insights into the nuanced mechanisms governing herbivore host plant selection (Hansen and Moran, 2014; Sugio et al., 2015).

Moving beyond the visual realm, olfactory cues, conveyed through volatile organic compounds (VOCs), serve as crucial elements in herbivore foraging. Microbes, including AMF and PGPR, alter plant volatiles, impacting herbivore attraction and below-ground interactions (Sharifi, Lee and Ryu, 2018; van Dijk, 2021). Similarly, pathogenic microbes influence olfactory cues, with vector-borne phytopathogens modifying plant VOCs to enhance vector attraction, thereby facilitating pathogen transmission (Grunseich et al., 2019). This complex interplay of plant-associated microbes, olfactory cues and herbivore behaviour contributes to a comprehensive understanding of the intricate dynamics that govern plant–pathogen–herbivore interactions. Furthermore, gustatory cues and non-volatile chemicals also play an important role in herbivore plant interaction. Plant microbes, such as AMF, influence plant nutrient acquisition, potentially modifying the host plant selection process (Das et al., 2022; Singh et al., 2022). In contrast, pathogenic microbes alter plant gustatory cues by changing defensive metabolites or nutritional quality, impacting herbivore preferences. Vector-borne phytopathogens also play a role by modifying gustatory cues to influence vector behaviour and enhance transmission success (Grunseich et al., 2019; Sardans et al., 2021). Understanding these microbial impacts on gustatory cues provides a holistic perspective on the multifaceted mechanisms that govern insect herbivore host plant selection in the presence of plant-associated microbes.

5.2.5 Change In Climate

The evolution of insect host plant choice is shaped by intricate factors, with theories such as "mother knows best" emphasising the link between oviposition preferences and offspring nutritional needs (Scheirs, Bruyn and Verhagen, 2000; Ryan, 2002; Calatayud et al., 2018). A meta-analysis reveals a strong correlation between host preference and performance in oligophagous species, while polyphagous species exhibit a lack of such correlation. Selection for host preference in polyphagous insects involves considerations beyond larval diet, including natural enemies, competition and the impact of pathogenic and beneficial organisms (Gripenberg et al., 2010). To overcome information limitations during host plant choice, polyphagous insects, especially females, exhibit higher behavioural pheno-typic plasticity in host selection (Carrasco, Larsson and Anderson, 2015). Previous experience with host plants also influence behaviour, influenced by plant-related cues during various life stages, play a role in host plant preference (Anderson and Anton, 2014). The intricate dynamics of these interactions highlight the need for further studies to uncover mechanisms governing experience-based plasticity in host preferences.

In the context of sensory aspects, the significance of signals facilitating interactions between plants and insects becomes prominent. Olfactory and chemosensory cues, vital for herbivorous insects, are shaped by selective pressures driving mutualistic or antagonistic interactions. While volatiles are crucial for plant recognition, the variation in volatile compounds within host plant species poses a challenge (Ninkovic, Markovic and Rensing, 2021). Polyphagous species may employ alternative strategies to extract fitness-related information across diverse host plant species (Anderson and Anton, 2014; Carrasco, Larsson and Anderson, 2015). The understanding of specific cues involved in host preferences and the sensory mechanisms governing experience-driven plasticity remains a focus of ongoing research. Advances in genetic studies shed light on chemosensory receptors, revealing the evolutionary relationships and functional loci linked to host attraction, oviposition and diversification of host races (Jones, 2007; Vertacnik and Linnen, 2017). The neural networks in insect brains, particularly the lateral horn and mushroom bodies, play pivotal roles in innate and plastic behavioural responses to plant volatiles (Carrasco, Larsson and Anderson, 2015). While studies have unravelled aspects of nectar-feeding neural pathways, mechanisms underlying experience-based plasticity in herbivorous insects remain largely unexplored. The integration of visual and olfactory cues in the central nervous system during foraging and oviposition decisions adds another layer of complexity, urging further research for a comprehensive understanding.

5.3 HOST PLANT SPECIALISATION VS. GENERALISATION

Host plant specialisation and generalisation are fundamental strategies employed by herbivorous insects in their complex process of host plant selection (Ali and Agrawal, 2012; Carrasco, Larsson and Anderson, 2015; Nobre et al., 2016). Factors influencing specialisation encompass a range of genetic, ecological and environmental determinants that guide herbivores toward specific plant species. Highly specialised herbivores have evolved to exploit the distinct chemical profiles of their preferred host plants, utilising specific chemical cues to locate and evaluate their food sources (Novotny et al., 2010; Kafle et al., 2014). In contrast, ecological and evolutionary implications arise from the trade-offs between specialisation and generalisation. Specialised herbivores often exhibit coevolutionary dynamics with their host plants, driving reciprocal adaptations. Generalists, however, display versatility in exploiting multiple plant species, potentially reducing their vulnerability to fluctuations in host availability (Poisot et al., 2011; García-Robledo and Horvitz, 2012). Within this context, the role of chemical cues cannot be overstated, as these cues play a pivotal role in shaping both the specialisation of herbivores and their ability to generalise across hosts. Chemical cues act as gatekeepers, determining whether an herbivore is attracted to, or repelled by, a potential host, thus underpinning the fine balance between specialisation and generalisation in herbivore–plant interactions (Wang et al., 2020).

5.4 EFFECT OF HOST PLANT QUALITY ON INSECT PERFORMANCE

Host plant quality, serving as a cornerstone in shaping the dynamics of insect fecundity and reproductive success, represents a multifaceted interplay between insects, predominantly herbivores, and their plant hosts. This intricate relationship extends beyond mere sustenance, encompassing shelter and oviposition, showcasing the depth of dependence between the two entities (Awmack and Leather, 2002; Moreau et al., 2017). The influence of host plant quality on insect performance is particularly emphasised in groups such as Lepidoptera, for which the quality of the host plant during larval development intricately links with both potential and achieved fecundity (Stern and Smith, 1960; Awmack and Leather, 2002). Coleoptera and aphids further exemplify the importance of host plant quality in fecundity, with challenging conditions prompting adaptive responses such as egg resorption to ensure survival, highlighting the direct impact of environmental factors on reproductive outcomes (Awmack and Leather, 2002; Ali et al., 2021; Makowe, 2023).

Moving beyond nutritional components, the direct effects of diet, specifically host plants, on insect fecundity are diverse and multifaceted. Nitrogen, a fundamental component, emerges as a key player in determining the reproductive success of herbivores, with aphids showcasing adaptations to modify nitrogen quality, maintaining high reproductive rates even on low-quality, low-nitrogen diets (Scriber, 1984; Leather, 2018). The interplay of carbon-based compounds, mineral nutrition and their intricate relationships further underscores the nuanced connection between nutritional factors, environmental conditions and herbivore adaptations (Awmack and Leather, 2002; Behmer, 2009; Whittaker, 2013). Defensive compounds within host plants, including nitrogen-containing compounds, cyanogenic glycosides, terpenoids and phenolics, introduce an additional layer of complexity to the host plant–insect interaction (Fürstenberg-Hägg, Zagrobelny and Bak, 2013). These compounds significantly impact the fecundity and performance of herbivorous insects. Notably, the cabbage looper and gypsy moths exemplify that these defensive compounds, rather than nutritional factors, play a key role in determining herbivore development (Gatehouse, 2002). Oviposition behaviour and achieved fecundity directly correlate with the presence of defensive components in host plants, as evidenced by the pine beauty moth and female sawflies (Awmack and Leather, 2002). The diverse effects observed across herbivore species, genetic adaptations and variations within a single plant underscore the intricate dynamics in these plant–insect interactions. Consequently, the quality of host plants emerges as a critical determinant influencing insect reproductive strategies (Kafle et al., 2014). In summary, the interplay between host plant quality and insect performance reflects the adaptability and trade-offs inherent in the complex ecological dynamics between plants and their herbivorous counterparts. The multifaceted nature of this relationship, encompassing nutritional components, defensive compounds and environmental factors, highlights the intricate web of interactions shaping the reproductive success of insects in their ecological niches.

5.5 IMPLICATIONS OF HOST PLANT SELECTION FOR PEST MANAGEMENT

Host plant resistance assumes a crucial role in pest management, particularly amidst the challenges posed by a projected global population of 10 billion by 2050. The imperative for increased food production is hindered by insect pests, causing substantial losses ranging from 25% to 30%. While chemical pesticides have historically been effective, their usage raises environmental concerns and contributes to resistance issues in pests (Kumari et al., 2022). In response, researchers advocate sustainable agriculture, emphasising host plant resistance as fundamental in diverse agro-ecosystems (Yang, 2008; Stout, 2014). Moreover, insect-resistant crop varieties, armed with antibiosis, antixenosis and tolerance mechanisms, emerge as an environmentally friendly alternative to chemical pesticides. Antibiosis influences pest biology, diminishing their population and damage, while antixenosis deters pests through non-preference for resistant plants (Heinrichs, 1985; Dwivedi and Pandit, 2024). Tolerance enables plants to resist or recover from pest-induced damage, establishing host plant resistance as a pivotal strategy for sustainable pest management. This alternative aligns with the imperative to balance food production with environmental responsibility (Smith, 2005; Kumari et al., 2022).

Considering host plant selection by herbivorous insects, there are profound implications for pest management strategies. The intricate interplay between chemical cues, herbivore behaviours and plant defences provides insights for developing effective and sustainable pest management approaches (Stout, 2014). Chemical ecology unveils the potential of semiochemicals for integrated pest management (IPM) strategies, such as "push-pull" approaches. Repellent semiochemicals drive pests away, while attractive ones guide them toward targeted control measures (Khan et al., 2016). This ecological approach minimises synthetic pesticide use while leveraging natural cues and behaviours to manage pest populations. Furthermore, understanding host plant selection guides the development of pest-resistant crop varieties, reducing reliance on chemical interventions and contributing to resilient, sustainable agricultural systems (Karlsson Green, Stenberg and Lankinen, 2020; Kumari et al., 2022). The multifaceted toolbox offered by host plant selection, within the context of chemical ecology, aligns with ecological principles, minimises environmental impacts and supports long-term agricultural productivity (Brzozowski and Mazourek, 2018). It facilitates the development of targeted pest control methods, using specific chemical cues to design traps and lures. Pheromone-based traps disrupt herbivore mating patterns, enhancing the precision and efficiency of pest control measures (Magalhães et al., 2018; Kansman et al., 2023). In essence, a deeper understanding of host plant selection offers a comprehensive approach to pest management, grounded in ecological principles and minimising environmental repercussions.

5.6 CONCLUSION

The intricate dance between herbivorous insects and their chosen host plants is a captivating example of the interplay between chemical ecology and ecological interactions. Throughout this chapter, we have explored the multifaceted world of host plant selection by herbivorous insects, driven by the intricate communication of chemical cues and signals. From the initial recognition of host plants through the detection of volatile compounds to the complex responses that shape feeding behaviours, this chapter has illuminated the dynamic processes that underpin this crucial ecological interaction. The chemical cues that guide herbivores toward their ideal host plants are not mere happenstance; they are the result of aeons of coevolution, adaptation and ecological pressures. As we delve into the complexities of herbivore behaviours, sensory mechanisms and the cascading effects on plant communities and ecosystems, it becomes abundantly clear that chemical ecology is an indispensable tool for understanding and managing herbivore populations. From the development of innovative pest control strategies that capitalise on semiochemicals to the design of pest-resistant crops, the practical applications of this knowledge have far-reaching implications for sustainable agriculture and ecosystem health. In closing, the intricate web of interactions between herbivorous insects and their host plants serves as a testament to the power of chemical communication in shaping the natural world. As we continue to unravel the mysteries of chemical ecology, this chapter stands as a testament to the profound insights that can be gained by exploring the intimate dialogues between organisms in their environment. By harnessing the wisdom of nature's chemical conversations, we pave the way for a more harmonious and balanced coexistence between insects and the plants that sustain them.

REFERENCES

Al-Khayri, J. M. *et al.* (2023) 'Plant secondary metabolites: the weapons for biotic stress management', *Metabolites*, 13(6), p. 716.

Ali, J. (2022) 'The chemical ecology of a model aphid pest, Myzus persicae, and its natural enemies'. Keele University.

Ali, J. (2023) 'The peach potato Aphid (Myzus persicae): ecology and management', 1, p. 132. doi: 10.1201/9781003400974.

Ali, J. G. and Agrawal, A. A. (2012) 'Specialist versus generalist insect herbivores and plant defense', *Trends in Plant Science*, 17(5), pp. 293–302.

Ali, M. Y. *et al.* (2021) 'Host-plant variations affect the biotic potential, survival, and population projection of Myzus persicae (Hemiptera: Aphididae)', *Insects*, 12(5), p. 375.

Anderson, P. and Anton, S. (2014) 'Experience-based modulation of behavioural responses to plant volatiles and other sensory cues in insect herbivores', *Plant, Cell & Environment*, 37(8), pp. 1826–1835.

Anton, S. and Cortesero, A.-M. (2022) 'Plasticity in chemical host plant recognition in herbivorous insects and its implication for pest control', *Biology*, 11(12), p. 1842.

Arimura, G. (2021) 'Making sense of the way plants sense herbivores', *Trends in Plant Science*, 26(3), pp. 288–298.

Awmack, C. S. and Leather, S. R. (2002) 'Host plant quality and fecundity in herbivorous insects', *Annual Review of Entomology*, 47(1), pp. 817–844.

Behmer, S. T. (2009) 'Insect herbivore nutrient regulation', *Annual Review of Entomology*, 54, pp. 165–187.

Blande, J. D., Holopainen, J. K. and Niinemets, Ü. (2014) 'Plant volatiles in polluted atmospheres: stress responses and signal degradation', *Plant, Cell & Environment*, 37(8), pp. 1892–1904.

Bruce, T. J. A. and Pickett, J. A. (2011) 'Perception of plant volatile blends by herbivorous insects-finding the right mix', *Phytochemistry*, 72(13), pp. 1605–1611. doi: 10.1016/j.phytochem.2011.04.011.

Brzozowski, L. and Mazourek, M. (2018) 'A sustainable agricultural future relies on the transition to organic agroecological pest management', *Sustainability*, 10(6), p. 2023.

Burkepile, D. E. and Parker, J. D. (2017) 'Recent advances in plant-herbivore interactions', *F1000Research*, 6.

Calatayud, P.-A. *et al.* (2018) 'Plant-insect interactions', *Ecology—Oxford Bibliographies*.

Carrasco, D., Larsson, M. C. and Anderson, P. (2015) 'Insect host plant selection in complex environments', *Current Opinion in Insect Science*, 8, pp. 1–7.

Chen, Y., Olson, D. M. and Ruberson, J. R. (2010) 'Effects of nitrogen fertilization on tritrophic interactions', *Arthropod-Plant Interactions*, 4, pp. 81–94.

Conchou, L. *et al.* (2019) 'Insect odorscapes: from plant volatiles to natural olfactory scenes', *Frontiers in Physiology*, 10, p. 972.

Dalin, P. *et al.* (2008) 'Leaf trichome formation and plant resistance to herbivory', *Induced Plant Resistance to Herbivory*, (Levin 1973), pp. 89–105. doi: 10.1007/978-1-4020-8182-8_4.

Das, P. P. *et al.* (2022) 'Plant-soil-microbes: a tripartite interaction for nutrient acquisition and better plant growth for sustainable agricultural practices', *Environmental Research*, 214, p. 113821.

van Dijk, L. J. A. (2021) 'Interactions between plants, microbes and insects'. Department of Ecology, Environment and Plant Sciences, Stockholm University.

Dwivedi, S. A. and Pandit, T. R. (2024) 'Host plant resistance and sustainable management of insect pests', *Antimicrobial resistance in agriculture and its consequences*. CRC Press, pp. 95–110.

Faiola, C. and Taipale, D. (2020) 'Impact of insect herbivory on plant stress volatile emissions from trees: a synthesis of quantitative measurements and recommendations for future research', *Atmospheric Environment: X*, 5, p. 100060.

Fernández, P. C. *et al.* (2019) 'The use of leaf surface contact cues during oviposition explains field preferences in the willow sawfly Nematus oligospilus', *Scientific Reports*, 9(1), p. 4946.

Fuentealba, A. *et al.* (2020) 'Leaf toughness as a mechanism of defence against spruce budworm', *Arthropod-Plant Interactions*, 14, pp. 481–489.

Fürstenberg-Hägg, J., Zagrobelny, M. and Bak, S. (2013) 'Plant defense against insect herbivores', *International Journal of Molecular Sciences*, 14(5), pp. 10242–10297.

Garcia-Robledo, C. and Horvitz, C. C. (2012) 'Jack of all trades masters novel host plants: positive genetic correlations in specialist and generalist insect herbivores expanding their diets to novel hosts', *Journal of Evolutionary Biology*, 25(1), pp. 38–53.

Gatehouse, J. A. (2002) 'Plant resistance towards insect herbivores: a dynamic interaction', *New Phytologist*, 156(2), pp. 145–169.

Gripenberg, S. *et al.* (2010) 'A meta-analysis of preference–performance relationships in phytophagous insects', *Ecology Letters*, 13(3), pp. 383–393.

Grunseich, J. M. *et al.* (2019) 'The role of plant-associated microbes in mediating host-plant selection by insect herbivores', *Plants*, 9(1), p. 6.

Hansen, A. K. and Moran, N. A. (2014) 'The impact of microbial symbionts on host plant utilization by herbivorous insects', *Molecular Ecology*, 23(6), pp. 1473–1496.

Heinrichs, E. A. (1985) *Genetic evaluation for insect resistance in rice*. International Rice Research Institute.

Jones, W. D. (2007) 'Insect host seeking: investigations into the molecular mechanisms of chemosensation'.

Kafle, D. *et al.* (2014) 'Genetic variation of the host plant species matters for interactions with above-and belowground herbivores', *Insects*, 5(3), pp. 651–667.

Kansman, J. T. *et al.* (2023) 'Chemical ecology in conservation biocontrol: new perspectives for plant protection', *Trends in Plant Science*, 28(10), pp. 1166–1177.

Karban, R. and Baldwin, I. T. (2007) *Induced responses to herbivory*. University of Chicago Press.

Karlsson Green, K., Stenberg, J. A. and Lankinen, Å. (2020) 'Making sense of Integrated Pest Management (IPM) in the light of evolution', *Evolutionary Applications*, 13(8), pp. 1791–1805.

Khan, Z. *et al.* (2016) 'Push-pull: chemical ecology-based integrated pest management technology', *Journal of Chemical Ecology*, 42, pp. 689–697.

Kitajima, K. *et al.* (2012) 'How cellulose-based leaf toughness and lamina density contribute to long leaf lifespans of shade-tolerant species', *New Phytologist*, 195(3), pp. 640–652.

Kumari, P. *et al.* (2022) 'Biotechnological approaches for host plant resistance to insect pests', *Frontiers in Genetics*, 13, p. 914029.

Leather, S. R. (2018) 'Factors affecting fecundity, fertility, oviposition, and larviposition in insects', *Insect reproduction*. CRC Press, pp. 143–174.

Levin, D. A. (1973) 'The role of trichomes in plant defense', *The Quarterly Review of Biology*, 48(1, Part 1), pp. 3–15.

Magalhães, D. M. *et al.* (2018) 'Identification of volatile compounds involved in host location by Anthonomus grandis (Coleoptera: Curculionidae)', *Frontiers in Ecology and Evolution*, 6, p. 98.

Makowe, I. A. (2023) 'An investigation on factors influencing the rearing and success of biological control agents of Gonipterus sp. n. 2 (Coleoptera: Curculionidae)'. University of Pretoria.

McCormick, A. C. *et al.* (2019) 'Herbivore-induced volatile emission from old-growth black poplar trees under field conditions', *Scientific Reports*, 9(1), pp. 1–10.

McCormick, A. C., Unsicker, S. B. and Gershenzon, J. (2012) 'The specificity of herbivore-induced plant volatiles in attracting herbivore enemies', *Trends in Plant Science*, 17(5), pp. 303–310.

Mechaber, W. L., Capaldo, C. T. and Hildebrand, J. G. (2002) 'Behavioral responses of adult female tobacco hornworms, Manduca sexta, to hostplant volatiles change with age and mating status', *Journal of Insect Science*, 2(1), p. 5.

Mello, M. O. and Silva-Filho, M. C. (2002) 'Plant-insect interactions: an evolutionary arms race between two distinct defense mechanisms', *Brazilian Journal of Plant Physiology*, 14(2), pp. 71–81.

Mithöfer, A. and Boland, W. (2012) 'Plant defense against herbivores: chemical aspects', *Annual Review of Plant Biology*, 63, pp. 431–450. doi: 10.1146/annurev-arplant-042110-103854.

Moreau, J. *et al.* (2017) 'How host plant and fluctuating environments affect insect reproductive strategies?', *Advances in botanical research*. Elsevier, pp. 259–287.

Mulatu, B. *et al.* (2006) 'Tomato fruit size, maturity and α-tomatine content influence the performance of larvae of potato tuber moth Phthorimaea operculella (Lepidoptera: Gelechiidae)', *Bulletin of Entomological Research*, 96(2), pp. 173–178.

Müller, C. and Riederer, M. (2005) 'Plant surface properties in chemical ecology', *Journal of Chemical Ecology*, 31, pp. 2621–2651.

Mumm, R., Posthumus, M. A. and Dicke, M. (2008) 'Significance of terpenoids in induced indirect plant defence against herbivorous arthropods', *Plant, Cell & Environment*, 31(4), pp. 575–585.

Ninkovic, V., Markovic, D. and Rensing, M. (2021) 'Plant volatiles as cues and signals in plant communication', *Plant, Cell & Environment*, 44(4), pp. 1030–1043.

Nobre, P. A. F. *et al.* (2016) 'Host-plant specialization mediates the influence of plant abundance on host use by flower head-feeding insects', *Environmental Entomology*, 45(1), pp. 171–177.

Novotny, V. *et al.* (2010) 'Guild-specific patterns of species richness and host specialization in plant–herbivore food webs from a tropical forest', *Journal of Animal Ecology*, 79(6), pp. 1193–1203.

Pacifico, D. *et al.* (2019) 'Caffeic acid and α-chaconine influence the resistance of potato tuber to Phthorimaea operculella (Lepidoptera: Gelechiidae)', *American Journal of Potato Research*, 96, pp. 403–413.

Pagare, S. *et al.* (2015) 'Secondary metabolites of plants and their role: overview', *Current Trends in Biotechnology and Pharmacy*, 9(3), pp. 293–304.

Parikh, G. L. *et al.* (2017) 'The influence of plant defensive chemicals, diet composition, and winter severity on the nutritional condition of a free-ranging, generalist herbivore', *Oikos*, 126(2).

Pickett, J. A. and Khan, Z. R. (2016) 'Plant volatile-mediated signalling and its application in agriculture: successes and challenges', *New Phytologist*, 212(4), pp. 856–870.

Poisot, T. *et al.* (2011) 'A conceptual framework for the evolution of ecological specialisation', *Ecology Letters*, 14(9), pp. 841–851.

Price, P. W. (1999) 'Host plant resource quality, insect herbivores and biocontrol', *Proceedings of the X international symposium on biological control of weeds*. US Department of Agriculture, Agricultural Research Service. Bozeman, MT, USA, pp. 583–5901.

Raj, V. P. *et al.* (2022) 'Plant-microbe-insect interactions: concepts and applications for agricultural sustainability', *Antifungal metabolites of rhizobacteria for sustainable agriculture*, p. 335.

Rani, P. U. (2015) 'Plant volatile chemicals and insect responses', *Plant biology and biotechnology: volume I: plant diversity, organization, function and improvement*, pp. 671–695.

Ryan, M. F. (2002) 'Plant chemicals', *Insect chemoreception: fundamental and applied*, pp. 27–72.

Sardans, J. *et al.* (2021) 'Ecometabolomics of plant–herbivore and plant–fungi interactions: a synthesis study', *Ecosphere*, 12(9), p. e03736.

Scheirs, J., Bruyn, L. De and Verhagen, R. (2000) 'Optimization of adult performance determines host choice in a grass miner', *Proceedings of the Royal Society of London. Series B: Biological Sciences*, 267(1457), pp. 2065–2069.

Scriber, J. M. (1984) 'Nitrogen nutrition of plants and insect invasion', *Nitrogen in crop production*, pp. 441–460.

Sharifi, R., Lee, S. and Ryu, C. (2018) 'Microbe-induced plant volatiles', *New Phytologist*, 220(3), pp. 684–691.

Shii, F. *et al.* (2021) 'Ultrasensitive detection by maxillary palp neurons allows non-host recognition without consumption of harmful allelochemicals', *Journal of Insect Physiology*, 132, p. 104263.

Shikano, I. *et al.* (2017) 'Tritrophic interactions: microbe-mediated plant effects on insect herbivores', *Annual Review of Phytopathology*, 55, pp. 313–331.

Singh, S. K. *et al.* (2022) 'Microbial enhancement of plant nutrient acquisition', *Stress Biology*, 2, pp. 1–14.

Smith, C. M. (2005) *Plant resistance to arthropods: molecular and conventional approaches*. Springer Science & Business Media.

Späthe, A. M. (2014) 'The function of volatile semiochemicals in host plant choice of ovipositing manduca moths (sphingidae)'. Jena, Friedrich-Schiller-Universität Jena, Diss., 2013.

Stern, V. M. and Smith, R. F. (1960) 'Factors affecting egg production and opposition in populations of Colias philodice eurytheme Boisduval (Lepidoptera: Pieridae)', *Hilgardia*, 29(10), pp. 411–454.

Stout, M. J. (2014) 'Host-plant resistance in pest management', *Integrated pest management*. Elsevier, pp. 1–21.

Sugio, A. *et al.* (2015) 'Plant–insect interactions under bacterial influence: ecological implications and underlying mechanisms', *Journal of Experimental Botany*, 66(2), pp. 467–478.

Tissier, A., Morgan, J. A. and Dudareva, N. (2017) 'Plant volatiles: going 'in'but not 'out'of trichome cavities', *Trends in Plant Science*, 22(11), pp. 930–938.

Turlings, T. C. J. and Erb, M. (2018) 'Tritrophic interactions mediated by herbivore-induced plant volatiles: mechanisms, ecological relevance, and application potential', *Annual Review of Entomology*, 63, pp. 433–452. doi: 10.1146/annurev-ento-020117-043507.

Vertacnik, K. L. and Linnen, C. R. (2017) 'Evolutionary genetics of host shifts in herbivorous insects: insights from the age of genomics', *Annals of the New York Academy of Sciences*, 1389(1), pp. 186–212.

Wang, D. *et al.* (2020) 'Plant chemistry determines host preference and performance of an invasive insect', *Frontiers in Plant Science*, 11, p. 594663.

Webster, B. and Cardé, R. T. (2017) 'Use of habitat odour by host-seeking insects', *Biological Reviews*, 92(2), pp. 1241–1249.

Wetzel, W. C. *et al.* (2023) 'Variability in plant–herbivore interactions', *Annual Review of Ecology, Evolution, and Systematics*, 54, pp. 451–474.

Whittaker, J. B. (2013) 'Impacts and responses at population level of herbivorous insects to elevated CO_2', *EJE*, 96(2), pp. 149–156.

Yadav, S. and Chattopadhyay, D. (2023) 'Lignin: the building block of defense responses to stress in plants', *Journal of Plant Growth Regulation*, 42, pp. 1–15.

Yang, L. (2008) *Integration of host plant resistance and biological control: using Arabidopsis-Insect interactions as a model system*. Wageningen University and Research.

CHAPTER 6

Ecology and Chemistry of Plant–Insect Interactions

6.1 INTRODUCTION

The complex and multifaceted relationships between plants and herbivores have captivated the attention of ecologists and chemists alike for decades. At the core of these interactions resides a complex interplay of ecological dynamics and chemical cues that influence the fundamental structure of ecosystems (Futuyma and Mitter, 1996; Sharma, Malthankar and Mathur, 2021). Understanding the profound influence of chemicals in mediating plant–herbivore interactions is not only a testament to the intricate design of nature but also holds the key to unlocking vital insights into the functioning of ecological systems (Mithöfer, Boland and Maffei, 2009; Nishida, 2014; Jayanthi, 2019). Plant–herbivore interactions, both in natural ecosystems and agricultural settings, are fundamentally mediated by chemical signals. From the emission of volatile compounds that serve as "scented" messengers between plants and their attackers to the synthesis of toxic secondary metabolites, the chemical cues exchanged between these players form the cornerstone of their dialogue (Meiners, 2015; Salazar, Jaramillo and Marquis, 2016). These cues serve as tools for communication, defence and negotiation in a continuous battle for survival and adaptation. The evolving arms race between plants and herbivores has

---

 DOI: 10.1201/9781003479857-6

led to a mesmerising array of chemical defence strategies that range from the production of toxic compounds to the induction of volatile emissions that recruit beneficial insects (Zu et al., 2020). This intricate chemistry underpins the ecological balance and biodiversity observed across various ecosystems.

In complex dynamics of ecological interactions, plant–herbivore relationships play a central role. Herbivores, ranging from tiny insects to large mammals, navigate through landscapes rich in chemical cues emitted by plants. These cues guide their foraging behaviours, dictate their host preferences and influence their population dynamics (Finch and Collier, 2000; Puente et al., 2008; Gitau et al., 2013). At the same time, plants have evolved an astonishing arsenal of chemical defences to fend off herbivore attacks (Engelberth et al., 2004; War et al., 2012, 2018). These defences create a delicate balance between consumption and protection, orchestrating a dance between attackers and defenders that reverberates through the trophic levels. In this chapter, we explore the chemistry pertaining to insect–plant interactions, encompassing insect performance (survival and fecundity), foraging and the selection of plants for shelter and ovipositional sites for egg laying. Furthermore, we investigate how herbivorous insects influence the performance of plants, manifesting in aspects such as nutritional status, damage and induced defence responses. This chapter aims to unravel the intricate connections inherent in nature, exploring how chemical cues impact herbivore behaviour and how plant responses to these interactions resonate across ecosystems. Through the perspective of chemical ecology, we uncover the concealed intricacies that regulate these relationships and gain insights into how they configure the diverse landscapes of our world. The exploration of the ecology and chemistry of plant–herbivore interactions not only provides us with an understanding of the mechanisms driving these connections but also yields a deeper insight into the fundamental interdependence of life on our planet.

6.2 HOW DO PLANTS DETECT PESTS?

Plants exhibit an extraordinary ability to perceive and react to various stages of herbivorous arthropod attacks. The process begins with the physical interaction between the herbivore and the plant, in which even subtle touches induced by wind or arthropod contact trigger a sensory response that extends across plant species. Subsequent phases, such as egg deposition, prompt plants to deploy direct and indirect defence mechanisms (Schaller and Weiler, 2002; Schaller, 2008; Hilker and Meiners,

2010). Plants employ intricate strategies, including the production of compounds to repel egg-laying females or induce ovicidal substances, as well
as feeding-induced responses that repel herbivores or modify the nutritive
value of their food (Karban and Myers, 1989; Karban and Baldwin, 2007;
Mostafa et al., 2022). Indirect defences, involving the release of extrafloral
nectar or plant volatiles, attract carnivorous enemies of herbivores and
serve as crucial elements in the complex web of ecological interactions
(Martin Heil, 2008; Heil, 2015; Jones, Koptur and von Wettberg, 2017).
Understanding how plants sense the three distinct phases – touch, oviposition and feeding – unravels the intricacies of their responses to herbivorous threats, offering insights into the interconnected nature of these
dynamic ecological relationships (Hilker and Meiners, 2010).

Touch, an important aspect of herbivorous arthropod attacks, involves
the activation of ion channels in the plasma membrane, translating physical damage into rapid changes in cytosolic Ca^{2+} concentrations (Mescher
and De Moraes, 2015; Gandhi et al., 2021). This cascade initiates signal
transduction pathways, leading to the induction of genes associated with
plant defence or growth (Jaffe, 1973; Braam and Davis, 1990; Braam, 2005).
For example, wheat responds to touch stimulation by upregulating the
transcription of lipoxygenase, a crucial enzyme in the octadecanoid pathway activated by damage (Mauch et al., 1997; Markovic et al., 2016). The
crawling of insect larvae induces the rapid synthesis of γ-aminobutyric
acid (GABA), showcasing the role of touch in indirect plant defence (Hall
et al., 2004). While the specificities of a plant's touch response to different species remain unclear, heightened GABA levels raise questions
about the molecular, physiological and ecological consequences of these
interactions.

Oviposition is an essential factor that assists plants in sensing insects.
During this process, the interaction between herbivorous arthropods and
plants becomes more complex. Ovipositional-induced plant changes triggers plant defensive responses, but it is the secretion released with egg
deposition that plays a pivotal role (Little et al., 2007; Reymond, 2013;
Lortzing et al., 2024). This phase involves highly specific responses tailored to both the plant and herbivore species, highlighting the intricate
chemical signalling and elicitors released during oviposition (Reymond,
2013). For instance, bruchins released by bruchid females induce direct
defence responses in plants, such as the growth of neoplasms and changes
in the transcription of several genes. Similarly, benzyl cyanide, released

with Pieris eggs during oviposition, triggers indirect defence mechanisms by causing alterations in the leaf surface. This, in turn, prompts egg parasitoids to intensify their search for host eggs and leads to the modification of transcription in numerous genes. Additionally, proteinaceous elicitors found in the secretion covering eggs play a crucial role in attracting egg parasitoids, thereby enhancing the plant's overall defence strategies (Doss et al., 1995; Doss, 2005; Fatouros et al., 2005). Ongoing studies aim to further explore characterising these oviposition-associated elicitors, shedding light on the sophisticated mechanisms underlying plant responses during this critical phase.

Another important factor that helps plants in sensing and detecting insect herbivores is feeding. During herbivore feeding, plants exhibit a remarkable ability to perceive not only the physical damage inflicted on their tissues but also specific cues released by the herbivores (Hilker and Meiners, 2010; Bonaventure, 2012). This intricate process involves the detection of nuanced signals, such as regurgitant deposited into plant wounds, containing compounds that play a crucial role in eliciting defensive responses. The plant responds dynamically to herbivore feeding by activating complex signalling pathways, involving changes in gene expression, the release of volatile compounds and the initiation of various defence-related processes (Bonaventure, 2012; Huffaker et al., 2013). Transitioning to the feeding phase, the complexity of plant responses extends beyond physical damage. Artificial wounding fails to replicate the full spectrum of responses, emphasising the significance of factors such as the extent and duration of inflicted damage (Rapo et al., 2019; Ali et al., 2024). Herbivores release regurgitant into plant wounds, acting as a crucial elicitor for defensive plant responses. Studies have identified specific transcriptional responses to elicitor compounds released by herbivorous attackers, showcasing the nuanced nature of plant recognition (Kudoh, Minamoto and Yamamoto, 2020). Elicitors such as volicitin, structurally related fatty acid-amino acid conjugates (FACs), caeliferins and inceptins induce a range of responses, including the production of defence-related phytohormones and the release of plant volatiles (Bonaventure, 2012, 2014; Aljbory and Chen, 2018; Abdul Malik, Kumar and Nadarajah, 2020). These findings enhance our understanding of the complex interaction between plants and herbivores during the feeding phase, elucidating the mechanisms through which plants detect and respond to herbivorous attacks.

6.3 INSECT FEEDING AND PLANT DEFENCE

The coevolutionary arms race between plants and herbivorous insects has sculpted the captivating diversity of defence strategies witnessed in the natural world (Kareiva, 1999; Fürstenberg-Hägg, Zagrobelny and Bak, 2013). Plants, in response to herbivore feeding patterns, have developed complex defence mechanisms connected to signalling pathways such as jasmonic acid (JA), salicylic acid (SA) and ethylene (ET) (War et al., 2012). Through these pathways, plants can deploy direct and indirect defence strategies, effectively deterring herbivorous insects (Kunkel and Brooks, 2002; Bari and Jones, 2009; Checker et al., 2018). *Brassica*, globally recognised as the second-largest oilseed crop after soybean, plays a vital role in addressing food security challenges (Attia et al., 2021). However, *Brassica* crops face significant annual losses due to insect pests, with approximately 50%–60% susceptibility to losses caused by plant pathogens originating from insects and mites (Poveda et al., 2020). The dependence on insecticides for pest management raises environmental concerns, emphasising the need for sustainable control strategies (Warwick, 2011; Baldwin et al., 2021). The plant immune system plays a crucial role in the dynamic interplay between plants and insect herbivores (Zhou and Zhang, 2020). Understanding the intricate molecular mechanisms of plant defence responses to varied feeding patterns of chewing and sucking insects is essential (War et al., 2018). Chewing and sucking behaviours trigger distinct plant responses, activating specific defence mechanisms (Walling, 2000).

6.4 DEFENCE RESPONSES IN BRASSICA AGAINST HERBIVOROUS INSECTS

Plants have been coexisting with and facing endless challenges from herbivorous insects for hundreds of millions of years. Plants, including *Brassica* species, have evolved an arsenal of defence strategies to combat herbivore attacks (Gatehouse, 2002; Ahuja et al., 2010). Plant defences are broadly classified as direct and indirect defences. Direct defences are plant traits (e.g., trichomes, secondary metabolites) that reduce their susceptibility to herbivorous insects or negatively affect insect biology or behaviour (Chen, 2008; War et al., 2012). Indirect defences are traits (e.g., herbivore-induced plant volatiles (HIPVs), extrafloral nectaries) that promote the attraction or efficacy of natural enemies of herbivorous insects such as predators and parasitoids (Heil, 2008; Aljbory and Chen, 2018). Both direct and indirect defences can be expressed constitutively (i.e., always present in plants) or induced following insect attack. The metabolic costs of induced defences are considered to be lower than constitutive defences, particularly when insect pressure is sporadic (Karban, 2011), and there could be a trade-off between constitutive and induced defence responses (Zhang et al., 2008). Plant phytohormone signalling networks, particularly JA and SA signalling pathways play crucial roles in optimising plant defences against herbivorous insects (Verma et al., 2016). In particular, the JA signalling cascade is considered a master regulator of induced plant responses to insect attack (Erb et al., 2012).

Brassica plants show a diverse array of direct physical and chemical defences against herbivorous insects. Among physical defences, epicuticular wax and trichomes account for one of the first lines of defences against herbivores. For example, the presence of epicuticular wax was found to enhance *Brassica oleracea* resistance to the diamondback moth (*Plutella xylostella*), flea beetles (*Phyllotreta* spp.) and cabbage stink bugs (*Eurydema* spp.) (Bohinc et al., 2014; Silva et al., 2017). Although such morphological structures are constitutive defences in *Brassica* plants, trichome density and epicuticular wax composition can be induced when challenged by insect herbivores (Traw and Dawson, 2002; Blenn et al., 2012).

The primary direct chemical defence in *Brassica* is the production of nitrogen- and sulphur-containing secondary metabolites known as glucosinolates (GS) that negatively affect insect herbivores (Hopkins et al., 2009; Jeschke et al., 2021), specifically generalist insects such as *Spodoptera littoralis* and *Mamestra brassicae* (Jeschke et al., 2017). GS are diverse in their structures (i.e., more than 130 known compounds) and are expressed constitutively in *Brassica* (Hopkins et al., 2009; Agerbirk and Olsen, 2012). The composition of GS in the family Brassicaceae varies depending on plant species, plant organs, ontogenetic stages, agricultural practices and environmental conditions (Textor and Gershenzon, 2009; Ahuja et al., 2010). Although GSs per se could be toxic to insects (Kim et al., 2008), they become highly toxic when hydrolysed by a specific enzyme called myrosinase and converted to toxic compounds such as isothiocyanates and nitriles (Agrawal and Kurashige, 2003; Wittstock et al., 2016). Both GSs and myrosinase are stored in adjacent but separate cells, and GSs only encounter the enzyme when plant tissues are mechanically damaged by insect feeding (Hopkins et al., 2009).

Even though Brassicaceous plants possess GSs constitutively, their levels, particularly that of indole GSs, in tissues can be induced rapidly and substantially following shoot or root herbivory by insects (Dam and Raaijmakers, 2006; Travers-Martin and Müller, 2007; Textor and Gershenzon, 2009). Insect attack can cause a redistribution of GSs in different organs or de novo synthesis of GSs in both attacked (i.e., local induction) and non-attacked (i.e., systemic induction) tissues (Hopkins et al., 2009; Touw et al., 2020). Likewise, the levels of myrosinase enzyme in plant tissues might increase upon insect feeding in some cases (Pontoppidan et al., 2003; Travers-Martin and Müller, 2007; Cachapa et al., 2021), although the impacts of such induction on plant defences remain uncertain (Textor and Gershenzon, 2009).

Considering that some specialist herbivores, such as *Pieris rapae* and *P. xylostella,* can neutralise GS (Ratzka et al., 2002; Wittstock et al., 2004), other secondary metabolites, such as phenolic compounds (e.g., flavonoids) and terpenoids (e.g., saponins), can confer direct resistance to specialist insects (Badenes-Perez et al., 2014; Ibrahim et al., 2018; Kovalikova et al., 2019). Moreover, cultivated *Brassica* plants can produce antioxidant defence enzymes such as polyphenol oxidase (PPO) and peroxidase (POD) and defensive proteins such as trypsin proteinase inhibitors (TPI) to defend specialist insects (Khattab, 2007; Ahmed et al., 2022). All these secondary metabolites and antioxidant enzymes can be present in *Brassica* constitutively or induced following insect attack or both (Ibrahim et al., 2018; Kovalikova et al., 2019).

Brassicaceous plants produce herbivore-induced plant volatiles (HIPVs) when attacked by pest herbivores, including glucosinolate breakdown products such as nitriles and isothiocyanates (Uefune et al., 2012; Mathur et al., 2013a; Zhou and Jander, 2022). The emission of HIPVs can deter insect herbivores (Verheggen et al., 2013) and attract their natural enemies, thus facilitating the top-down control of herbivorous insects (Puente et al., 2008; Mathur et al., 2013a). Furthermore, *Brassica juncea* can produce extrafloral nectaries as an indirect defence, which can be present in plants constitutively, but the amount of nectar production could be induced following insect feeding (Mathur et al., 2013b). The possession and induction of such nectaries could support natural enemies of herbivores by providing alternative foods (Jamont et al., 2013; Mathur et al., 2013).

6.5 INSECT-MEDIATED IMPACTS ON PLANTS

Delving into the symbiotic dance between plants and herbivorous insects, this section explores the multifaceted impacts of herbivory, viral transmission and honeydew on plant ecosystems. From the tangible effects of direct insect damage to the subtle intricacies of induced defences and the implications of honeydew in fostering plant diseases, this exploration unveils the delicate equilibrium defining plant–insect relationships.

6.5.1 Insect Damage and Its Impact on Plant Chemistry

Herbivory, the act of insect consumption of plant tissues, goes beyond mere physical harm (Hodkinson, 2012; Myers and Sarfraz, 2017). The consumptive act reduces the plant's photosynthetic capacity, affecting resource allocation for growth and reproduction (Kerchev et al., 2012; Zhou et al., 2015). Mechanical damage, whether through chewing, boring or scraping, reshapes the plant's architecture, influencing its resilience to environmental stressors (Wright and Vincent, 1996; Ali et al., 2024). Oviposition injury further amplifies direct damage, as insect eggs deposited on or within plant tissues cause localised harm (Little et al., 2007; Hilker and Fatouros, 2015). These interactions leave an indelible mark on plant structural integrity and overall health. However, the repercussions of herbivorous insect activity extend beyond the visible. Plants respond to such attacks by unleashing a complex arsenal of defence mechanisms. Signalling compounds, such as herbivore-induced plant volatiles and green leaf volatiles, emanate from broken or punctured leaf tissues (War et al., 2011). Chemicals such as GSs, initially stored in an inactive state, activate upon damage, altering the plant's nutritional quality and defence strategies (Bruce, 2014; Singh, 2017).

Herbivorous insects release unique compounds, triggering gall formation in plants or inducing changes that lead to the formation of physical barriers in leaves (Nakamura, Miyamoto and Ohgushi, 2003; Raman, 2011; Raman, Fantasia and Fagan, 2012). This chemical warfare further influences the attractiveness of plants to other herbivores and predators, sculpting the broader ecological landscape. Insect-triggered modifications extend beyond the aforementioned aspects, exerting a discernible influence on resource allocation within plant systems (Orians, Thorn and Gómez, 2011; Züst and Agrawal, 2017). In response to herbivory, plants recalibrate growth and invest resources in developing defence strategies. These strategies span the production of antibiotics, antifeedants, deterrents and defensive volatile compounds. (War et al., 2011). Moreover, insect damage orchestrates the release of volatile compounds acting as signalling molecules. These compounds convey crucial information to neighbouring undamaged plants, creating an ecological symphony that communicates the presence of infested host plants (Ninkovic et al., 2013;

Conrath et al., 2015). Adding another layer to these intricate dynamics, herbivorous insects secret compounds during feeding that suppress the plant's defence system (Ali, 2023; Ali, Bayram, et al., 2023). As we explore the intricate world of insect damage and its profound chemical ramifications, we unravel the interconnected mechanisms shaping the complex dynamics between plants and herbivores.

6.5.2 Insect Secretion (Honeydew) and Plant Interaction

Honeydew, a sugary excretion produced by phloem-feeding insects, emerges as a key player (Álvarez-Pérez, Lievens and de Vega, 2023). As insects tap into plant sap for sustenance, they leave a sugary trail in the form of honeydew (Dhami et al., 2011). This viscous substance, often overlooked, acts as a mediator, influencing the dynamics between insects and plants. The composition and quantity of honeydew are shaped by various ecological factors, including the insect species involved, the nutritional quality of the plant sap and environmental conditions (Fischer and Shingleton, 2001; Schillewaert et al., 2017; Blanchard et al., 2022). Recognising honeydew as a crucial aspect of insect–plant relationships opens the door to unravelling its intricate contributions to ecological processes (Tena et al., 2016, 2018; Álvarez-Pérez, Lievens and de Vega, 2023). Beyond its role as excretion, honeydew holds ecological significance that extends to various facets of the ecosystem (Buitenhuis et al., 2004; Tena et al., 2016; Colazza, Peri and Cusumano, 2023) (Figure 6.1). This sticky substance, rich in sugars and other compounds, serves as a nutrient source for a diverse array of microorganisms (Tena et al., 2016). The microbial communities thriving within honeydew, in turn, contribute to the broader ecological landscape (Owen and Wiegert, 1976; Leroy et al., 2011; van Neerbos et al., 2020). The VOCs released by these microorganisms can influence plant health, pest behaviour and even attract natural enemies of insects (Álvarez-Pérez, Lievens and de Vega, 2023). As we explore the ecological consequences of honeydew, it becomes evident that this seemingly inconspicuous substance has far-reaching implications for biodiversity, nutrient cycling and pest management (Álvarez-Pérez, Lievens and de Vega, 2023).

6.6 POSITIVE INSECT–PLANT INTERACTION

Insect–plant interactions are not solely characterised by negativity; a significant aspect involves positive interactions that play a crucial role in agriculture and ecosystems (Wielkopolan and Obrępalska-Stęplowska, 2016; Pincebourde et al., 2017; Sharma, Malthankar and Mathur, 2021).

How do plants 'DETECT' insect attack?

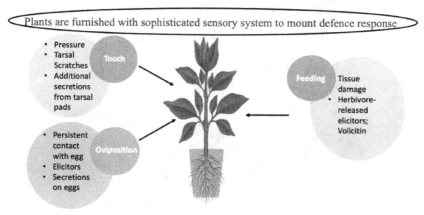

FIGURE 6.1   Illustration showing how plants utilise touch, oviposition and feeding cues to detect herbivores. Mechanoreceptors sense physical stimuli caused by herbivore movement (1), while the recognition of oviposition cues (2) triggers pre-emptive defence responses. Feeding activity (3) induces chemical signalling that activates plant defences and alerts neighbouring plants through the release of volatiles, facilitating coordinated antiherbivore strategies.

Insects engaging in these interactions are deemed beneficial, falling broadly into categories such as natural enemies (predators, parasitoids, entomopathogens) and pollinators. The relationships formed by these insects, particularly predators, parasitoids and pollinators, are mutualistic in nature (Nicholls and Altieri, 2013; Wielkopolan and Obrępalska-Stęplowska, 2016; Brodeur et al., 2017). They offer protection to plants, and in return, plants provide them with essential resources. For instance, plants produce extra floral nectar, attracting natural enemies that contribute to safeguarding the plants from herbivorous insects (Figure 6.2) (Brodeur et al., 2017). Similarly, pollinator insects visit plants for sustenance, and their interactions aid in the crucial process of pollination (Nicholls and Altieri, 2013; Getanjaly, Sharma and Kushwaha, 2015). These intricate relationships, complemented by visual cues, are primarily orchestrated by chemical communication between plants and insects (Zu et al., 2023). Such chemicals manifest in the nutritional composition of plants and the volatile compounds released, adapting to contextual needs. For instance, when plants face biotic stress, they release a specific blend of volatiles that attracts natural enemies (Turlings, Tumlinson and Lewis, 1990; Ninkovic

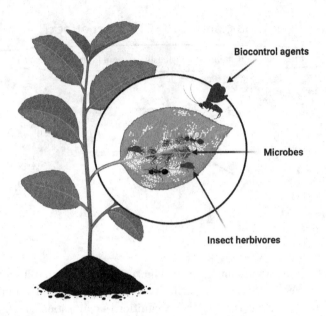

Biocontrol agents

Microbes

Insect herbivores

FIGURE 6.2 Illustration showing the multifaceted interactions involving honey-dew deposition on a plant leaf. The central circle showcases honeydew accumulation, a critical element in the intricate relationships between plants, honeydew, microbes, insect herbivores and biocontrol agents.

et al., 2013; Xiu et al., 2019). These phytochemical cues become instrumental guides, assisting natural enemies in locating their host.

6.7 CHEMICAL SIGNALLING IN PLANT–HERBIVORE INTERACTIONS

The complex dynamics between plants and herbivores is choreographed by an array of chemical cues that traverse the boundaries of sight and sound, forming an intricate language of their own. VOCs, allelochemicals and pheromones serve as the communicative tools in this dynamic exchange, conveying information that guides the behaviours and actions of both parties (El-Shafie and Faleiro, 2017). These chemical signals are not mere byproducts of metabolic processes; they are the nuanced whispers that plants and herbivores use to communicate, negotiate and navigate their interactions. VOCs, for instance, are released by plants in response to herbivore attacks, acting as distress signals that alert neighbouring plants to impending danger (Kost and Heil, 2006; Ninkovic et al., 2013). These volatiles not only communicate the presence of herbivores but also provide information about the nature of the attack, allowing neighbouring plants to activate their own defence mechanisms. Allelochemicals, another category of chemical cues, are bioactive compounds that plants release into their surroundings to influence the growth and behaviour of other organisms, including herbivores. These allelochemicals can serve as defensive chemicals by deterring herbivores from feeding on the plant, thereby reducing the extent of damage (Hickman et al., 2021). Additionally, pheromones play a crucial role in the communication between herbivores of the same species. These chemical messengers allow herbivores to convey information about their presence, mating status and even optimal feeding sites to conspecifics (Howse, Stevens and Jones, 1998; Cardé and Millar, 2009; Shorey, 2013) (see Chapter 2 for details). Thus, chemical signals serve as the invisible threads that connect plants and herbivores in a dynamic web of interactions, fostering a symphony of responses that reverberate throughout the ecosystem.

The language of chemical cues is an essential tool in the communication toolbox of both plants and herbivores, allowing them to interact even in the absence of direct physical contact. For instance, when a herbivore inflicts damage upon a plant, the plant responds by emitting a volatile blend that not only alerts neighbouring plants to the threat but also summons the help of natural enemies, such as parasitoids and predators, that prey on the herbivores (Puente et al., 2008; War et al., 2011). This indirect defence strategy, known as "plant-mediated indirect defence," illustrates the far-reaching consequences of chemical signalling in shaping the dynamics of plant–herbivore interactions (Arimura et al., 2004; Yuan et al., 2008). Furthermore, herbivores have evolved to detect and interpret these chemical cues to their advantage. For example, many herbivores use plant volatiles to locate suitable host plants or to assess the presence of competitors and predators (Dicke, 2000; Robert et al., 2012). In turn, herbivores emit their own pheromones to communicate with conspecifics, guiding them toward optimal feeding sites or away from areas with high predation risk. This intricate web of communication underscores the profound role of chemical signalling in influencing the behaviours, movements and interactions of herbivores within the ecological landscape. Overall, the language of chemicals forms an integral part of the dialogue between plants and herbivores, shaping the course of their interactions and reverberating throughout the intricate tapestry of ecosystems.

6.8 HERBIVORE ADAPTATIONS TO PLANT CHEMICAL DEFENCES

In response to plant defence against herbivory, herbivores employ various adaptations to navigate and thrive amidst plant chemical defences (Mithöfer and Boland, 2012; War et al., 2012, 2020). One primary strategy involves avoidance, wherein herbivores modify their feeding behaviour to steer clear of toxic compounds present in certain plants. Additionally, herbivores may employ mechanisms to suppress the plant's defence responses, ensuring a more favourable feeding environment (Alba et al., 2011; Ibanez, Gallet and Després, 2012). Another noteworthy adaptation involves the production of detoxifying enzymes by herbivores. These enzymes play a crucial role in neutralising or breaking down toxic compounds present in the ingested plant material, allowing herbivores to consume a wider range of plants (Ahmad, 1986; Birnbaum and Abbot, 2018). Furthermore, herbivores showcase adaptability through the acquisition of symbiotic microbes. These microbes aid in the detoxification process, providing herbivores with a microbial support system that enhances their ability to handle and neutralise plant toxins effectively (Hansen and Moran, 2014; Hammer and Bowers, 2015; Itoh et al., 2018). Furthermore, another fascinating aspect of herbivore adaptations is the occurrence of mutations in phytochemical target sites (Ali, 2023). This genetic variability allows herbivores to evolve and develop resistance to specific plant defences, further highlighting their remarkable ability to coexist with and adapt to the chemical intricacies of their plant hosts (Rotter and Holeski, 2018; Ali, 2023).

6.9 IMPLEMENTATION OF CHEMISTRY OF INSECT–PLANT INTERACTION

Understanding the intricate dynamics of insect–plant interactions, particularly the nuanced interplay between ecology and chemistry, is paramount for devising sustainable pest control strategies and safeguarding crop productivity (Meiners, 2015; Cortesero et al., 2016). At the core of this interaction lies the exchange of phytochemicals, shaping the susceptibility or resistance levels of host plants (Figure 6.3) (Mithöfer, Boland and Maffei, 2009; Ali, Wei, et al., 2023). Identifying plant-derived chemicals and comprehending how herbivorous insects respond to these compounds not only aids in deciphering the host plant's resilience but also unveils insights into the development of insect attractants and deterrents (Dyer et al., 2018). This

Insect-Plant Herbivore Interactions

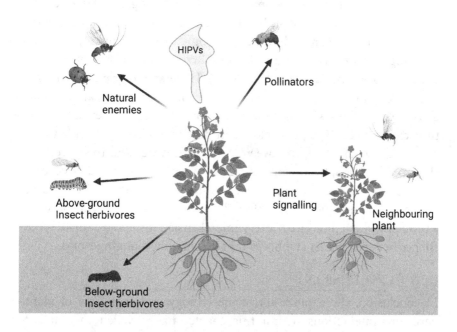

FIGURE 6.3 Illustration showing insect–plant interactions: a complex network including pollinators, natural enemies, above-ground and below-ground insect herbivores, herbivore-released plant volatiles and plant signalling inducing defence in neighbouring plants.

knowledge, in turn, opens avenues for discerning the signalling pathways in plants responsible for the production or suppression of these chemicals, offering genetic information that can be harnessed to develop resistant crop varieties (Bruce, Wadhams and Woodcock, 2005; Divekar et al., 2022). Exploring the further impact of these chemicals on insect behaviour, we uncover the potential to develop pheromone traps disrupting insect herbivore mating cycles. Simultaneously, for beneficial insects, this knowledge facilitates the establishment of traps and cards, enhancing the recruitment of biocontrol agents in the field (Wang et al., 2022). As the field of chemical ecology advances, researchers are presented with exciting opportunities to unravel new mysteries. Cutting-edge technologies, including advanced mass spectrometry and genomic approaches, enable the identification and quantification of intricate chemical compounds. Integrating omics technologies with ecological and behavioural studies promises a holistic understanding of the multifaceted interactions between plants and herbivores (Rossignol et al., 2006; Mbaluto et al., 2020; Roychowdhury et al., 2023). The transformative applications of insights gleaned from chemical ecology research are vast. In sustainable agriculture, leveraging plant chemical defences can lead to the development of resilient crop varieties, reducing reliance on chemical pesticides (Brzozowski and Mazourek, 2018; Farooq and Pisante, 2019; Muhie, 2022). Understanding plant–herbivore interactions informs precision pest management strategies, allowing targeted interventions with minimal environmental impact. Beyond agriculture, chemical ecology principles contribute to biodiversity conservation, unravelling relationships between plants, herbivores and their predators (Khan et al., 2008; Schuman et al., 2016). This knowledge guides ecosystem restoration and the protection of endangered species. The fusion of emerging technologies with profound chemical ecology insights promises a future in which molecular-level comprehension fosters sustainable coexistence between plants and herbivores in a rapidly changing world.

6.10 CONCLUSION

In conclusion, the exploration of the ecology and chemistry of plant–herbivore interactions within this chapter has unveiled the intricate mechanisms governing this dynamic relationship. Beginning with an understanding of how plants detect pests, we delved into the multifaceted realm of insect feeding and plant defence, particularly highlighting the defenceresponses exhibited by *Brassica* against herbivorous insects. The far-reaching impacts of insect-mediated interactions on plants were elucidated, encompassing the repercussions of insect damage on plant chemistry, the nuanced dynamics of insect secretions and their interaction with plants and the often-overlooked positive aspects of insect–plant interactions. The pivotal role of chemical signalling in orchestrating these interactions was underscored, laying the foundation for a comprehensive comprehension of the intricate web of relationships. Finally, an exploration of herbivore adaptations to plant chemical defences and the practical implementation of the chemistry of insect–plant interactions shed light on the broader ecological implications of these phenomena. This chapter serves as a holistic guide, providing a nuanced perspective on the dynamic interplay between plants and herbivores, emphasising the pivotal role of chemistry in shaping their ecological relationships.

REFERENCES

Abdul Malik, N. A., Kumar, I. S. and Nadarajah, K. (2020) 'Elicitor and receptor molecules: Orchestrators of plant defense and immunity', *International Journal of Molecular Sciences*, 21(3), p. 963. doi: 10.3390/ijms21030963.

Agerbirk, N. and Olsen, C. E. (2012) 'Glucosinolate structures in evolution', *Phytochemistry*, 77, pp. 16–45.

Agrawal, A. A. and Kurashige, N. S. (2003) 'A role for isothiocyanates in plant resistance against the specialist herbivore Pieris rapae', *Journal of Chemical Ecology*, 29, pp. 1403–1415.

Ahmad, S. (1986) 'Enzymatic adaptations of herbivorous insects and mites to phytochemicals', *Journal of Chemical Ecology*, 12, pp. 533–560.

Ahmed, M. A. *et al.* (2022) 'Oviposition preference and two-sex life table of Plutella xylostella and its association with defensive enzymes in three Brassicaceae crops', *Crop Protection*, 151, p. 105816.

Ahuja, I., Rohloff, J. and Bones, A. M. (2010) 'Defence mechanisms of Brassicaceae: implications for plant-insect interactions and potential for integrated pest management. A review', *Agronomy for Sustainable Development*, 30(2), pp. 311–348. doi: 10.1051/agro/2009025.

Alba, J. M. *et al.* (2011) 'Avoidance and suppression of plant defenses by herbivores and pathogens', *Journal of Plant Interactions*, 6(4), pp. 221–227.

Ali, J. (2023) 'The peach potato aphid (Myzus persicae): ecology and management', 1, p. 132. doi: 10.1201/9781003400974.

Ali, J., Bayram, A., *et al.* (2023) 'Peach–potato aphid Myzus persicae: current management strategies, challenges, and proposed solutions', *Sustainability*, 15(14), p. 11150.

Ali, J., Wei, D., *et al.* (2023) 'Exogenous application of methyl salicylate induces defence in brassica against peach potato Aphid Myzus persicae', *Plants*, 12(9), p. 1770.

Ali, J. *et al.* (2024) 'Wound to survive: mechanical damage suppresses aphid performance on brassica', *Journal of Plant Diseases and Protection*, 131, pp. 1–12.

Aljbory, Z. and Chen, M. (2018) 'Indirect plant defense against insect herbivores: a review', *Insect Science*, 25(1), pp. 2–23.

Álvarez-Pérez, S., Lievens, B. and de Vega, C. (2023) 'Floral nectar and honeydew microbial diversity and their role in biocontrol of insect pests and pollination', *Current Opinion in Insect Science*, 61, p. 101138.

Arimura, G. I. *et al.* (2004) 'Herbivore-induced defense response in a model legume. Two-spotted spider mites induce emission of (E)-β-ocimene and transcript accumulation of (E)-β-ocimene synthase in Lotus japonicus', *Plant Physiology*, 135(4), pp. 1976–1983. doi: 10.1104/pp.104.042929.

Attia, Z. *et al.* (2021) 'Breeding for sustainable oilseed crop yield and quality in a changing climate', *Theoretical and Applied Genetics*, 134(6), pp. 1817–1827.

Badenes-Perez, F. R., Gershenzon, J. and Heckel, D. G. (2014) 'Insect attraction versus plant defense: young leaves high in glucosinolates stimulate oviposition by a specialist herbivore despite poor larval survival due to high saponin content', *PLoS One*, 9(4), p. e95766.

Baldwin, J. M. *et al.* (2021) 'Occurrence of arthropod pests associated with Brassica carinata and impact of defoliation on yield', *GCB Bioenergy*, 13(4), pp. 570–581.

Bari, R. and Jones, J. D. G. (2009) 'Role of plant hormones in plant defence responses', *Plant Molecular Biology*, 69, pp. 473–488.

Birnbaum, S. S. L. and Abbot, P. (2018) 'Insect adaptations toward plant toxins in milkweed–herbivores systems–a review', *Entomologia Experimentalis et Applicata*, 166(5), pp. 357–366.

Blanchard, S. *et al.* (2022) 'Combined elevation of temperature and CO2 impacts the production and sugar composition of aphid honeydew', *Journal of Chemical Ecology*, 48(9–10), pp. 772–781.

Blenn, B. *et al.* (2012) 'Insect egg deposition induces indirect defense and epicuticular wax changes in Arabidopsis thaliana', *Journal of Chemical Ecology*, 38, pp. 882–892.

Bohinc, T., Marković, D. and Trdan, S. (2014) 'Leaf epicuticular wax as a factor of antixenotic resistance of cabbage to cabbage flea beetles and cabbage stink bugs attack', *Acta Agriculturae Scandinavica, Section B—Soil & Plant Science*, 64(6), pp. 493–500.

Bonaventure, G. (2012) 'Perception of insect feeding by plants', *Plant Biology*, 14(6), pp. 872–880.

Bonaventure, G. (2014) 'Plants recognize herbivorous insects by complex signalling networks', *Annual Plant Reviews: Insect-Plant Interactions*, 47, pp. 1–35.

Braam, J. (2005) 'In touch: plant responses to mechanical stimuli', *New Phytologist*, 165(2), pp. 373–389. doi: 10.1111/j.1469-8137.2004.01263.x.

Braam, J. and Davis, R. W. (1990) 'Rain-, wind-, and touch-induced expression of calmodulin and calmodulin-related genes in Arabidopsis', *Cell*, 60(3), pp. 357–364.

Brodeur, J. *et al.* (2017) 'Predators, parasitoids and pathogens', *Aphids as crop pests*. CABI Wallingford UK, pp. 225–261.

Bruce, T. J. A. (2014) 'Glucosinolates in oilseed rape: secondary metabolites that influence interactions with herbivores and their natural enemies', *Annals of Applied Biology*, 164(3), pp. 348–353.

Bruce, T. J. A., Wadhams, L. J. and Woodcock, C. M. (2005) 'Insect host location: a volatile situation', *Trends in Plant Science*, 10(6), pp. 269–274.

Brzozowski, L. and Mazourek, M. (2018) 'A sustainable agricultural future relies on the transition to organic agroecological pest management', *Sustainability*, 10(6), p. 2023.

Buitenhuis, R. *et al.* (2004) 'The role of honeydew in host searching of aphid hyperparasitoids', *Journal of Chemical Ecology*, 30, pp. 273–285.

Cachapa, J. C. *et al.* (2021) 'Induction and priming of plant defense by root-associated insect-pathogenic fungi', *Journal of Chemical Ecology*, 47(1), pp. 112–122.

Cardé, R. T. and Millar, J. G. (2009) 'Pheromones', *Encyclopedia of insects*. Elsevier, pp. 766–772.

Checker, V. G. *et al.* (2018) 'Role of phytohormones in plant defense: signaling and cross talk', *Molecular aspects of plant-pathogen interaction*, pp. 159–184.

Chen, M. S. (2008) 'Inducible direct plant defense against insect herbivores: a review', *Insect Science*, 15(2), pp. 101–114. doi: 10.1111/j.1744-7917.2008.00190.x.

Colazza, S., Peri, E. and Cusumano, A. (2023) 'Chemical ecology of floral resources in conservation biological control', *Annual Review of Entomology*, 68, pp. 13–29.

Conrath, U. *et al.* (2015) 'Priming for enhanced defense', *Annual Review of Phytopathology*, 53(1), pp. 97–119. doi: 10.1146/annurev-phyto-080614-120132.

Cortesero, A. *et al.* (2016) 'Chemical ecology: an integrative and experimental science', *Chemical ecology*, pp. 23–46.

Dam, N. M. van and Raaijmakers, C. E. (2006) 'Local and systemic induced responses to cabbage root fly larvae (Delia radicum) in Brassica nigra and B. oleracea', *Chemoecology*, 16, pp. 17–24.

Dhami, M. K. *et al.* (2011) 'Species-specific chemical signatures in scale insect honeydew', *Journal of Chemical Ecology*, 37, pp. 1231–1241.

Dicke, M. (2000) 'Chemical ecology of host-plant selection by herbivorous arthropods: a multitrophic perspective', *Biochemical Systematics and Ecology*, 28(7), pp. 601–617.

Divekar, P. A. *et al.* (2022) 'Plant secondary metabolites as defense tools against herbivores for sustainable crop protection', *International Journal of Molecular Sciences*, 23(5), p. 2690.

Doss, R. P. (2005) 'Treatment of pea pods with Bruchin B results in up-regulation of a gene similar to MtN19', *Plant Physiology and Biochemistry*, 43(3), pp. 225–231.

Doss, R. P. *et al.* (1995) 'Response of Np mutant of pea (Pisum sativum L.) to pea weevil (Bruchus pisorum L.) oviposition and extracts', *Journal of Chemical Ecology*, 21, pp. 97–106.

Dyer, L. A. *et al.* (2018) 'Modern approaches to study plant–insect interactions in chemical ecology', *Nature Reviews Chemistry*, 2(6), pp. 50–64.

El-Shafie, H. A. F. and Faleiro, J. R. (2017) 'Semiochemicals and their potential use in pest management', *Biological control of pest and vector insects*, pp. 10–5772.

Engelberth, J. *et al.* (2004) 'Airborne signals prime plants against insect herbivore attack', *Proceedings of the National Academy of Sciences of the United States of America*, 101(6), pp. 1781–1785. doi: 10.1073/pnas.0308037100.

Erb, M., Meldau, S. and Howe, G. A. (2012) 'Role of phytohormones in insect-specific plant reactions', *Trends in Plant Science*, 17(5), pp. 250–259. doi: 10.1016/j.tplants.2012.01.003.

Farooq, M. and Pisante, M. (2019) *Innovations in sustainable agriculture*. Springer.

Fatouros, N. E. *et al.* (2005) 'Oviposition-induced plant cues: do they arrest Trichogramma wasps during host location?', *Entomologia Experimentalis et Applicata*, 115(1), pp. 207–215.

Finch, S. and Collier, R. H. (2000) 'Host-plant selection by insects–a theory based on "appropriate/inappropriate landings" by pest insects of cruciferous plants', *Entomologia experimentalis et applicata*, 96(2), pp. 91–102.

Fischer, M. K. and Shingleton, A. W. (2001) 'Host plant and ants influence the honeydew sugar composition of aphids', *Functional Ecology*, 15(4), pp. 544–550.

Fürstenberg-Hägg, J., Zagrobelny, M. and Bak, S. (2013) 'Plant defense against insect herbivores', *International Journal of Molecular Sciences*, 14(5), pp. 10242–10297.

Futuyma, D. J. and Mitter, C. (1996) 'Insect—plant interactions: the evolution of component communities', *Philosophical Transactions of the Royal Society of London. Series B: Biological Sciences*, 351(1345), pp. 1361–1366.

Gandhi, A. *et al.* (2021) 'Deciphering the role of ion channels in early defense signaling against herbivorous insects', *Cells*, 10(9), p. 2219.

Gatehouse, J. A. (2002) 'Plant resistance towards insect herbivores: a dynamic interaction', *New Phytologist*, 156(2), pp. 145–169.

Getanjaly, V. L. R., Sharma, P. and Kushwaha, R. (2015) 'Beneficial insects and their value to agriculture', *Research Journal of Agriculture and Forestry Sciences ISSN*, 2320, p. 6063.

Gitau, C. W. *et al.* (2013) 'A review of semiochemicals associated with bark beetle (Coleoptera: Curculionidae: Scolytinae) pests of coniferous trees: a focus on beetle interactions with other pests and their associates', *Forest Ecology and Management*, 297, pp. 1–14. doi: 10.1016/j.foreco.2013.02.019.

Hall, D. E. *et al.* (2004) 'Footsteps from insect larvae damage leaf surfaces and initiate rapid responses', *European Journal of Plant Pathology*, 110, pp. 441–447.

Hammer, T. J. and Bowers, M. D. (2015) 'Gut microbes may facilitate insect herbivory of chemically defended plants', *Oecologia*, 179, pp. 1–14.

Hansen, A. K. and Moran, N. A. (2014) 'The impact of microbial symbionts on host plant utilization by herbivorous insects', *Molecular Ecology*, 23(6), pp. 1473–1496.

Heil, M. (2008) 'Indirect defence via tritrophic interactions', *New Phytologist*, 178(1), pp. 41–61. doi: 10.1111/j.1469-8137.2007.02330.x.

Heil, M. (2015) 'Extrafloral nectar at the plant-insect interface: a spotlight on chemical ecology, phenotypic plasticity, and food webs', *Annual Review of Entomology*, 60, pp. 213–232.

Hickman, D. T. *et al.* (2021) 'Allelochemicals as multi-kingdom plant defence compounds: towards an integrated approach', *Pest Management Science*, 77(3), pp. 1121–1131.

Hilker, M. and Fatouros, N. E. (2015) 'Plant responses to insect egg deposition', *Annual Review of Entomology*, 60, pp. 493–515.

Hilker, M. and Meiners, T. (2010) 'How do plants "notice" attack by herbivorous arthropods?', *Biological Reviews*, 85(2), pp. 267–280. doi: 10.1111/j.1469-185X.2009.00100.x.

Hodkinson, I. (2012) *Insect herbivory*. Springer Science & Business Media.

Hopkins, R. J., van Dam, N. M. and van Loon, J. J. A. (2009) 'Role of glucosinolates in insect-plant relationships and multitrophic interactions', *Annual Review of Entomology*, 54, pp. 57–83.

Howse, P. E., Stevens, I. D. R. and Jones, O. T. (1998) 'Insect semiochemicals and communication', *Insect pheromones and their use in pest management*. Springer, pp. 3–37.

Huffaker, A. *et al.* (2013) 'Plant elicitor peptides are conserved signals regulating direct and indirect antiherbivore defense', *Proceedings of the National Academy of Sciences*, 110(14), pp. 5707–5712.

Ibanez, S., Gallet, C. and Després, L. (2012) 'Plant insecticidal toxins in ecological networks', *Toxins*, 4(4), pp. 228–243.

Ibrahim, S. *et al.* (2018) 'Herbivore and phytohormone induced defensive response in kale against cabbage butterfly, Pieris brassicae Linn', *Journal of Asia-Pacific Entomology*, 21(1), pp. 367–373.

Itoh, H. *et al.* (2018) 'Detoxifying symbiosis: microbe-mediated detoxification of phytotoxins and pesticides in insects', *Natural Product Reports*, 35(5), pp. 434–454.

Jaffe, M. J. (1973) 'Thigmomorphogenesis: the response of plant growth and development to mechanical stimulation: with special reference to Bryonia dioica', *Planta*, 114, pp. 143–157.

Jamont, M., Piva, G. and Fustec, J. (2013) 'Sharing N resources in the early growth of rapeseed intercropped with faba bean: does N transfer matter?', *Plant and Soil*, 371, pp. 641–653.

Jayanthi, P. D. (2019) 'Ecological chemistry of insect-plant interactions', *Journal of Eco-friendly Agriculture*, 14(2), pp. 1–10.

Jeschke, V. *et al.* (2017) 'How glucosinolates affect generalist lepidopteran larvae: growth, development and glucosinolate metabolism', *Frontiers in Plant Science*, 8, p. 1995.

Jeschke, V. *et al.* (2021) 'So much for glucosinolates: a generalist does survive and develop on Brassicas, but at what cost?', *Plants*, 10(5), p. 962.

Jones, I. M., Koptur, S. and von Wettberg, E. J. (2017) 'The use of extrafloral nectar in pest management: overcoming context dependence', *Journal of Applied Ecology*, 54(2), pp. 489–499.

Karban, R. (2011) 'The ecology and evolution of induced resistance against herbivores', *Functional Ecology*, 25(2), pp. 339–347. doi: 10.1111/j.1365-2435.2010.01789.x.

Karban, R. and Baldwin, I. T. (2007) *Induced responses to herbivory*. University of Chicago Press.

Karban, R. and Myers, J. H. (1989) 'Induced plant responses to herbivory', *Annual Review of Ecology and Systematics*, 20(1), pp. 331–348. doi: 10.1146/annurev .es.20.110189.001555.

Kareiva, P. (1999) 'Coevolutionary arms races: is victory possible?', *Proceedings of the National Academy of Sciences*, 96(1), pp. 8–10.

Kerchev, P. I. *et al.* (2012) 'Plant responses to insect herbivory: interactions between photosynthesis, reactive oxygen species and hormonal signalling pathways', *Plant, Cell & Environment*, 35(2), pp. 441–453.

Khan, Z. R. *et al.* (2008) 'Chemical ecology and conservation biological control', *Biological Control*, 45(2), pp. 210–224.

Khattab, H. (2007) 'The defense mechanism of cabbage plant against phloem-sucking aphid (Brevicoryne brassicae L.)', *Australian Journal of Basic and Applied Sciences*, 1(1), pp. 56–62.

Kim, J. H. *et al.* (2008) 'Identification of indole glucosinolate breakdown products with antifeedant effects on Myzus persicae (green peach aphid)', *The Plant Journal*, 54(6), pp. 1015–1026.

Kost, C. and Heil, M. (2006) 'Herbivore-induced plant volatiles induce an indirect defence in neighbouring plants', *Journal of Ecology*, 94(3), pp. 619–628.

Kovalikova, Z. *et al.* (2019) 'Changes in content of polyphenols and ascorbic acid in leaves of white cabbage after pest infestation', *Molecules*, 24(14), p. 2622.

Kudoh, A., Minamoto, T. and Yamamoto, S. (2020) 'Detection of herbivory: eDNA detection from feeding marks on leaves', *Environmental DNA*, 2(4), pp. 627–634.

Kunkel, B. N. and Brooks, D. M. (2002) 'Cross talk between signaling pathways in pathogen defense', *Current Opinion in Plant Biology*, 5(4), pp. 325–331.

Leroy, P. D. *et al.* (2011) 'Microorganisms from aphid honeydew attract and enhance the efficacy of natural enemies', *Nature Communications*, 2(1), p. 348.

Little, D. *et al.* (2007) 'Oviposition by pierid butterflies triggers defense responses in Arabidopsis', *Plant Physiology*, 143(2), pp. 784–800.

Lortzing, V. *et al.* (2024) 'Plant defensive responses to insect eggs are inducible by general egg-associated elicitors', *Scientific Reports*, 14(1), p. 1076.

Markovic, D. *et al.* (2016) 'Plant responses to brief touching: a mechanism for early neighbour detection?', *PLoS One*, 11(11), p. e0165742.

Mathur, V., Tytgat, T. O. G., *et al.* (2013a) 'An ecogenomic analysis of herbivore-induced plant volatiles in Brassica juncea', *Molecular Ecology*, 22(24), pp. 6179–6196.

Mathur, V., Wagenaar, R., *et al.* (2013b) 'A novel indirect defence in Brassicaceae: structure and function of extrafloral nectaries in Brassica juncea', *Plant, Cell & Environment*, 36(3), pp. 528–541.

Mauch, F. *et al.* (1997) 'Mechanosensitive expression of a lipoxygenase gene in wheat', *Plant Physiology*, 114(4), pp. 1561–1566.

Mbaluto, C. M. *et al.* (2020) 'Insect chemical ecology: chemically mediated interactions and novel applications in agriculture', *Arthropod-plant Interactions*, 14, pp. 671–684.

Meiners, T. (2015) 'Chemical ecology and evolution of plant–insect interactions: a multitrophic perspective', *Current Opinion in Insect Science*, 8, pp. 22–28.

Mescher, M. C. and De Moraes, C. M. (2015) 'Role of plant sensory perception in plant–animal interactions', *Journal of Experimental Botany*, 66(2), pp. 425–433.

Mithöfer, A. and Boland, W. (2012) 'Plant defense against herbivores: chemical aspects', *Annual Review of Plant Biology*, 63, pp. 431–450. doi: 10.1146/annurev-arplant-042110-103854.

Mithöfer, A., Boland, W. and Maffei, M. E. (2009) 'Chemical ecology of plant-insect interactions', *Molecular aspects of plant disease resistance*. Wiley-Blackwell, pp. 261–291.

Mostafa, S. *et al.* (2022) 'Plant responses to herbivory, wounding, and infection', *International Journal of Molecular Sciences*, 23(13), p. 7031.

Muhie, S. H. (2022) 'Novel approaches and practices to sustainable agriculture', *Journal of Agriculture and Food Research*, 10, p. 100446.

Myers, J. H. and Sarfraz, R. M. (2017) 'Impacts of insect herbivores on plant populations', *Annual Review of Entomology*, 62, pp. 207–230.

Nakamura, M., Miyamoto, Y. and Ohgushi, T. (2003) 'Gall initiation enhances the availability of food resources for herbivorous insects', *Functional Ecology*, 17, pp. 851–857.

van Neerbos, F. A. C. *et al.* (2020) 'Honeydew composition and its effect on life-history parameters of hyperparasitoids', *Ecological Entomology*, 45(2), pp. 278–289.

Nicholls, C. I. and Altieri, M. A. (2013) 'Plant biodiversity enhances bees and other insect pollinators in agroecosystems. A review', *Agronomy for Sustainable Development*, 33, pp. 257–274.

Ninkovic, V. *et al.* (2013) 'Volatile exchange between undamaged plants - a new mechanism affecting insect orientation in intercropping', *PloS One*, 8(7), p. e69431. doi: 10.1371/journal.pone.0069431.

Nishida, R. (2014) 'Chemical ecology of insect–plant interactions: ecological significance of plant secondary metabolites', *Bioscience, Biotechnology, and Biochemistry*, 78(1), pp. 1–13.

Orians, C. M., Thorn, A. and Gómez, S. (2011) 'Herbivore-induced resource sequestration in plants: why bother?', *Oecologia*, 167, pp. 1–9.

Owen, D. F. and Wiegert, R. G. (1976) 'Do consumers maximize plant fitness?', *Oikos*, 27, pp. 488–492.

Pincebourde, S. *et al.* (2017) 'Plant–insect interactions in a changing world', *Advances in botanical research*. Elsevier, pp. 289–332.

Pontoppidan, B. *et al.* (2003) 'Infestation by cabbage aphid (Brevicoryne brassicae) on oilseed rape (Brassica napus) causes a long lasting induction of the myrosinase system', *Entomologia Experimentalis et Applicata*, 109(1), pp. 55–62.

Poveda, J. *et al.* (2020) 'Development of transgenic Brassica crops against biotic stresses caused by pathogens and arthropod pests', *Plants*, 9(12), p. 1664.

Puente, M. *et al.* (2008) 'Impact of herbivore-induced plant volatiles on parasitoid foraging success: a spatial simulation of the Cotesia rubecula, Pieris rapae, and Brassica oleracea system', *Journal of Chemical Ecology*, 34, pp. 959–970.

Raman, A. (2011) 'Morphogenesis of insect-induced plant galls: facts and questions', *Flora-Morphology, Distribution, Functional Ecology of Plants*, 206(6), pp. 517–533.

Raman, V., Fantasia, R. and Fagan, J. M. (2012) 'Pesticides and decline in pollinator populations'.

Rapo, C. B. *et al.* (2019) 'Feeding intensity of insect herbivores is associated more closely with key metabolite profiles than phylogenetic relatedness of their potential hosts', *PeerJ*, 7, p. e8203.

Ratzka, A. *et al.* (2002) 'Disarming the mustard oil bomb', *Proceedings of the National Academy of Sciences*, 99(17), pp. 11223–11228.

Reymond, P. (2013) 'Perception, signaling and molecular basis of oviposition-mediated plant responses', *Planta*, 238, pp. 247–258. doi: 10.1007/s00425-013-1908-y.

Robert, C. A. M. *et al.* (2012) 'Herbivore-induced plant volatiles mediate host selection by a root herbivore', *New Phytologist*, 194(4), pp. 1061–1069.

Rossignol, M. *et al.* (2006) 'Plant proteome analysis: a 2004–2006 update', *Proteomics*, 6(20), pp. 5529–5548.

Rotter, M. C. and Holeski, L. M. (2018) 'A meta-analysis of the evolution of increased competitive ability hypothesis: genetic-based trait variation and herbivory resistance trade-offs', *Biological Invasions*, 20, pp. 2647–2660.

Roychowdhury, R. *et al.* (2023) 'Multi-omics pipeline and omics-integration approach to decipher plant's abiotic stress tolerance responses', *Genes*, 14(6), p. 1281.

Salazar, D., Jaramillo, A. and Marquis, R. J. (2016) 'The impact of plant chemical diversity on plant–herbivore interactions at the community level', *Oecologia*, 181, pp. 1199–1208.

Schaller, A. (2008) *Induced plant resistance to herbivory*. Springer.

Schaller, F. and Weiler, E. W. (2002) 'Wound-and mechanical signalling', *Plant signal transduction*. Oxford University Press, pp. 20–44.

Schillewaert, S. *et al.* (2017) 'The effect of host plants on genotype variability in fitness and honeydew composition of Aphis fabae', *Insect Science*, 24(5), pp. 781–788.

Schuman, M. C. *et al.* (2016) 'How does plant chemical diversity contribute to biodiversity at higher trophic levels?', *Current Opinion in Insect Science*, 14, pp. 46–55.

Sharma, G., Malthankar, P. A. and Mathur, V. (2021) 'Insect–plant interactions: a multilayered relationship', *Annals of the Entomological Society of America*, 114(1), pp. 1–16.

Shorey, H. H. (2013) *Animal communication by pheromones*. Academic Press.

Silva, G. A. *et al.* (2017) 'Wax removal and diamondback moth performance in collards cultivars', *Neotropical Entomology*, 46, pp. 571–577.

Singh, A. (2017) 'Glucosinolates and plant defense', *Glucosinolates*. Springer International Publishing, pp. 237–246.

Tena, A. *et al.* (2016) 'Parasitoid nutritional ecology in a community context: the importance of honeydew and implications for biological control', *Current Opinion in Insect Science*, 14, pp. 100–104.

Tena, A. *et al.* (2018) 'The influence of aphid-produced honeydew on parasitoid fitness and nutritional state: a comparative study', *Basic and Applied Ecology*, 29, pp. 55–68.

Textor, S. and Gershenzon, J. (2009) 'Herbivore induction of the glucosinolate–myrosinase defense system: major trends, biochemical bases and ecological significance', *Phytochemistry Reviews*, 8, pp. 149–170.

Touw, A. J. *et al.* (2020) 'Both biosynthesis and transport are involved in glucosinolate accumulation during root-herbivory in Brassica rapa', *Frontiers in Plant Science*, 10, p. 1653.

Travers-Martin, N. and Müller, C. (2007) 'Specificity of induction responses in Sinapis alba L. and their effects on a specialist herbivore', *Journal of Chemical Ecology*, 33, pp. 1582–1597.

Traw, B. M. and Dawson, T. E. (2002) 'Differential induction of trichomes by three herbivores of black mustard', *Oecologia*, 131, pp. 526–532.

Turlings, T. C. J., Tumlinson, J. H. and Lewis, W. J. (1990) 'Exploitation of herbivore-induced plant odors by host-seeking parasitic wasps', *Science*, 250(4985), pp. 1251–1253. doi: 10.1126/science.250.4985.1251.

Uefune, M. *et al.* (2012) 'Herbivore-induced plant volatiles enhance the ability of parasitic wasps to find hosts on a plant', *Journal of Applied Entomology*, 136(1–2), pp. 133–138.

Verheggen, F. J. *et al.* (2013) 'Aphid responses to volatile cues from turnip plants (Brassica rapa) infested with phloem-feeding and chewing herbivores', *Arthropod-Plant Interactions*, 7, pp. 567–577.

Verma, V., Ravindran, P. and Kumar, P. P. (2016) 'Plant hormone-mediated regulation of stress responses', *BMC Plant Biology*, 16, pp. 1–10.

Walling, L. L. (2000) 'The myriad plant responses to herbivores', *Journal of Plant Growth Regulation*, 19, pp. 195–216.

Wang, H.-L. *et al.* (2022) 'Insect pest management with sex pheromone precursors from engineered oilseed plants', *Nature Sustainability*, 5(11), pp. 981–990.

War, A. R. *et al.* (2011) 'Herbivore induced plant volatiles: their role in plant defense for pest management', *Plant Signaling and Behavior*, 6(12), pp. 1973–1978. doi: 10.4161/psb.6.12.18053.

War, A. R. *et al.* (2012) 'Mechanisms of plant defense against insect herbivores', *Plant Signaling and Behavior*, 7(10). doi: 10.4161/psb.21663.

War, A. R. *et al.* (2018) 'Plant defence against herbivory and insect adaptations', *AoB Plants*, 10(4), p. ply037.

War, A. R. *et al.* (2020) 'Plant defense and insect adaptation with reference to secondary metabolites', *Co-evolution of secondary metabolites*, pp. 795–822.

Warwick, S. I. (2011) 'Brassicaceae in agriculture', *Genetics and genomics of the Brassicaceae*, pp. 33–65.

Wielkopolan, B. and Obrępalska-Stęplowska, A. (2016) 'Three-way interaction among plants, bacteria, and coleopteran insects', *Planta*, 244(2), pp. 313–332.

Wittstock, U. *et al.* (2004) 'Successful herbivore attack due to metabolic diversion of a plant chemical defense', *Proceedings of the National Academy of Sciences*, 101(14), pp. 4859–4864.

Wittstock, U. *et al.* (2016) 'Glucosinolate breakdown', *Advances in botanical research*. Elsevier, pp. 125–169.

Wright, W. and Vincent, J. F. V. (1996) 'Herbivory and the mechanics of fracture in plants', 71(3), pp. 401–413. https://doi.org/10.1111/j.1469-185X.1996.tb01280.x

Xiu, C. *et al.* (2019) 'Herbivore-induced plant volatiles enhance field-level parasitism of the mirid bug Apolygus lucorum', *Biological Control*, 135, pp. 41–47.

Yuan, J. S. *et al.* (2008) 'Elucidation of the genomic basis of indirect plant defense against insects', *Plant Signaling and Behavior*, 3(9), pp. 720–721. doi: 10.4161/psb.3.9.6468.

Zhang, P.-J. *et al.* (2008) 'Trade-offs between constitutive and induced resistance in wild crucifers shown by a natural, but not an artificial, elicitor', *Oecologia*, 157, pp. 83–92.

Zhou, J. M. and Zhang, Y. (2020) 'Plant immunity: danger perception and signaling', *Cell*, 181(5), pp. 978–989. doi: 10.1016/j.cell.2020.04.028.

Zhou, S. *et al.* (2015) 'Alteration of plant primary metabolism in response to insect herbivory', *Plant Physiology*, 169(3), pp. 1488–1498.

Zhou, S. and Jander, G. (2022) 'Molecular ecology of plant volatiles in interactions with insect herbivores', *Journal of Experimental Botany*, 73(2), pp. 449–462.

Zu, P. *et al.* (2020) 'Information arms race explains plant-herbivore chemical communication in ecological communities', *Science*, 368(6497), pp. 1377–1381.

Zu, P. *et al.* (2023) 'Plant–insect chemical communication in ecological communities: an information theory perspective', *Journal of Systematics and Evolution*, 61(3), pp. 445–453.

Züst, T. and Agrawal, A. A. (2017) 'Trade-offs between plant growth and defense against insect herbivory: an emerging mechanistic synthesis', *Annual Review of Plant Biology*, 68(1), pp. 513–534.

CHAPTER 7

Chemical Ecology and Agriculture

7.1 INTRODUCTION

Chemical ecology and agriculture constitute a dynamic and interconnected discipline, providing insights into the relationships between plants, insects and the chemical cues that govern their interactions. Insects take advantage of chemical compounds, exploiting agricultural crops and yields for their benefit (Sondheimer, 2012; Mbaluto et al., 2020). Chemical ecology plays a crucial role in shaping the agricultural landscape, exploring both its negative and positive impacts (Dicke, 2000; Johnson and Gregory, 2006; Webster and Cardé, 2017; Kansman et al., 2023). In terms of negative effects, chemical ecology reveals a complex narrative in which herbivorous insects navigate their environment using chemical communication, focusing on plants as sources of sustenance, shelter and oviposition (Webster and Cardé, 2017). The high precision with which herbivorous insects distinguish between host and non-host plants, guided by chemical cues released by the vegetation, underscores the intricate dynamics between the two (Bruce and Pickett, 2011; Murphy and Loewy, 2015). This chemical interplay, however, exacts a toll on agriculture, leading to substantial losses in production. As herbivorous insects feed, the release of chemicals serves to numb the feeding site in plants, making it inconspicuous and suppressing the plant's defence mechanisms. The consequence is a delicate balance, governed by the chemical exchange between plants and

DOI: 10.1201/9781003479857-7

insects, with significant implications for agriculture production (Bos et al., 2010; Hogenhout and Bos, 2011; Zhang et al., 2013). Conversely, the positive impact of chemical ecology on agriculture emerges through the collection and identification of phytochemicals, laying the foundation for understanding insect behaviour in response to specific plant-produced compounds (Webster et al., 2008; Pickett et al., 2012). This knowledge extends to the development of innovative pest management strategies, including traps, lures and mating disruption approaches. These strategies, informed by synthetic analogues of plant chemicals, not only reduce pest populations and damage but also foster the recruitment and biodiversity of biocontrol agents in agricultural fields (Renwick, 2002; Reddy and Guerrero, 2010; Mbaluto et al., 2020; Kansman et al., 2023).

Moreover, phytochemicals play a pivotal role in activating plant defence through exogenous applications of compounds such as salicylic acid and jasmonic acid derivatives (War et al., 2020). This induced defence mechanism, coupled with plant priming, contributes to enhanced stress tolerance and improved agricultural yields (Lou et al., 2005; Walters, Newton and Lyon, 2014; Ali, Wei, et al., 2023). Beyond defence, phytochemicals emerge as key players in promoting plant growth, offering protection against both biotic and abiotic stresses (Pieterse et al., 2012; Chen and Pang, 2023). Throughout this chapter, we will explore the multifaceted contributions of chemical ecology to agriculture, unravelling its role in mitigating herbivorous insect damage, enhancing plant growth and fostering sustainable agricultural practices. From reducing toxic insecticidal burdens to activating plant defences and implementing innovative pest management strategies, chemical ecology stands as a cornerstone in the quest for resilient and productive agricultural systems.

7.2 INSECT HERBIVORES AND HOST PLANT CHEMICALS

In the complex interplay of chemical ecology and agriculture, the role of herbivorous insects is profound. These creatures, particularly herbivores, display a remarkable ability to exploit the chemical cues emitted by plants, influencing their foraging behaviour and significantly impacting agriculture (Mbaluto et al., 2020). Insect herbivores, driven by chemical communication, exhibit a precise ability to identify and target specific crops for feeding. For instance, the Colorado potato beetle (*Leptinotarsa decemlineata*) relies on volatile compounds released by potato plants to locate its preferred host (Cingel et al., 2016). This targeted approach can result in substantial crop damage, leading to significant yield losses for farmers and economic ramifications for agriculture. Beyond direct feeding, herbivorous insects employ chemical signals such as pheromones to communicate and aggregate (Wyatt, 2014; Pirk, 2021). The western corn rootworm (*Diabrotica virgifera*), through the release of aggregation pheromones, can attract conspecifics to join in feeding on a specific host plant (Choe, 2019; Brzozowski et al., 2020). This aggregative behaviour intensifies the impact on agriculture by concentrating herbivore pressure on particular crops.

Continuous exposure to chemical cues from host plants can drive the evolution of chemical resistance in herbivores. This adaptability poses an ongoing challenge for agriculture, as herbivores may develop mechanisms to circumvent plant defences or become less susceptible to traditional pest control methods (Després, David and Gallet, 2007; War et al., 2020). The chemical cues used by herbivores not only affect host plants but can also indirectly impact beneficial insects. For example, predatory insects that rely on plant-emitted volatiles to locate prey may face challenges in finding herbivorous insects when these prey species utilise chemical strategies to evade detection. This disruption in natural predator–prey dynamics can contribute to imbalances in insect populations within agricultural ecosystems. In summary, the complex chemical interactions between herbivorous insects and host plants have multifaceted negative implications for agriculture. These encompass direct feeding damage, aggregative behaviours, the evolution of resistance and disruptions to beneficial insect dynamics, highlighting the need for a comprehensive understanding of these interactions for effective pest management in agriculture (Yew and Chung, 2015; Erb and Reymond, 2019; Abd El-Ghany, 2020).

Phytotoxins are chemical compounds produced by plants that affect the growth, development or physiology of other organisms. In the context of plant pathology, phytotoxins play important roles in plant defences against pathogens and pests as well as in the development and spread of plant diseases (Graniti, 1991). Phytotoxins can directly or indirectly inhibit the growth and development of plant pathogens, suppressing disease symptoms and spread. For example, some phytotoxins inhibit fungal growth, disrupt pathogen cell membranes or interfere with pathogen enzymes, thus preventing or reducing infection (Graniti, 1991; Tuerkkan and Dolar, 2008). On the other hand, some phytotoxins can also promote disease development. Some plant pathogens, such as bacteria and fungi, produce toxins that damage plant cells and tissues, causing disease symptoms such as wilting, necrosis and discoloration. These toxins can also trigger plant defence responses, such as the production of reactive oxygen species and phytohormones, which can further exacerbate the disease symptoms and spread (Andolfi et al., 2011; Uoman et al., 2018) Moreover, some phytotoxins play a role in plant–microbe interactions by modulating the plant's immune system and influencing the composition of the microbial community associated with the plant. Some compounds, such as phytoalexins, can activate plant defences and inhibit pathogen growth, while others can act as signalling molecules that attract beneficial microbes to the plant (Graniti, 1991). Overall, phytotoxins are complex compounds with both positive and negative roles in plant pathology, depending on the context and the specific interactions involved. Understanding the mechanisms and functions of phytotoxins can help in developing strategies to control plant diseases and improve crop resilience.

7.3 PHYTOCHEMICAL AND PLANT GROWTH

Phytohormones play a pivotal role in regulating plant growth and development, broadly categorised into activators and inhibitors (Sosnowski, Truba and Vasileva, 2023). Activators such as auxin, gibberellin and cytokinin promote cell division, fruiting and seed formation, enhancing overall plant growth. Conversely, inhibitors such as abscisic acid and ethylene can impede growth or trigger abscission, with ethylene displaying dual roles based on context (Márquez, Alarcón and Salguero, 2016; Egamberdieva et al., 2017). An extensive body of literature showed that endogenous and exogenous application of these phytohormones enhances plant growth (Cabello-Conejo, Prieto-Fernández and Kidd, 2014; Tantasawat, Sorntip and Pornbungkerd, 2015; Ahmad et al., 2019; Kosakivska et al., 2022).

7.3.1 Auxin

Auxin, the first-discovered phytohormone, holds particular significance due to its involvement in nearly all plant processes (Woodward and Bartel, 2005; Wu et al., 2011). Synthesised in shoot apical meristems, young leaves, seeds and fruits, auxin plays a pivotal role in cell division. Auxins, including indole-3-acetic acid (IAA), indole-3-butyric acid (IBA) and 4-chloroindole-3-acetic acid (4-CL-IAA), represent the first-discovered plant hormones and hold a central role in promoting plant growth and development (Sachs, 2005; Márquez, Alarcón and Salguero, 2016). They are closely linked to root and shoot formation, exerting regulatory control over general plant growth. The concentration of auxins within plants is intricately regulated, involving influx and efflux from tissues, biosynthesis from tryptophan and the formation of IAA conjugates (Tanimoto, 2005).

Auxin's influence extends across numerous plant processes, both directly and indirectly, and it collaborates with other phytohormones (Robles et al., 2012; Paque and Weijers, 2016). These hormone interactions impact cell expansion, cell cycle progression, leaf development and embryonic development during seed maturation (Jurado et al., 2010).

In conjunction with ethylene, auxins contribute to the regulation of primary root elongation by modulating cell proliferation in the root apical meristem and cell elongation in the elongation zone (Street et al., 2015). Ethylene plays a crucial role in auxin regulation, affecting the transcription of auxin biosynthesis and transport genes. During ethylene treatment, auxin accumulates in root tips, highlighting the intricate balance between these hormones (Méndez-Bravo et al., 2019). Imbalances in ethylene levels can lead to elevated auxin content, inhibiting cell elongation and primary root growth (Negi et al., 2010). Genes such as auxin resistant 2 (AXR2) and AXR3 are key components of the auxin signalling pathway, and their loss of function also reduces ethylene sensitivity in primary roots (Stepanova et al., 2007; Swarup et al., 2007).

Moreover, auxins play a role in mitigating plant stress responses, enhancing tolerance to stressors such as heavy metals. For instance, the addition of IAA improved the growth of lead-affected sunflower plants, leading to increased root volume and surface area diameter (Fässler et al., 2010). Additionally, auxin levels can decrease in rice plants subjected to salinity stress, potentially triggering the activation of other phytohormones such as abscisic acid (ABA) to influence plant growth (Iqbal and Ashraf, 2013). Microbes, such as plant growth-promoting rhizosphere (PGPR) bacteria, produce auxins that enhance plant growth. Utilising bacteria strains with antifungal properties, growth in horticultural plants such as *Physalis ixocarpa* was increased by 45%, attributed to bacterial auxin production (Méndez-Bravo et al., 2023). These microbe-derived auxins present a sustainable alternative to chemical fertilisers, as they can stimulate various metabolic processes, ultimately improving different aspects of plant growth (Ahmed and Hasnain, 2014).

7.3.2 Gibberellins

Gibberellins (GAs) form a group of approximately 136 identified compounds, although not all are directly involved in plant development (Hedden, 2020). GAs play pivotal roles in various aspects of plant growth, including seed germination, leaf expansion, flower initiation, stem elongation and fruit development (Yamaguchi, 2008). Studies, such as one by

Khan et al. (2014) on tomato plants, have revealed substantial enhancements in fruit yield, leaf area, shoot length and root dry weight following inoculation with endophytic bacteria *Sphingomonas* sp. Notably, the presence of physiologically active GAs produced by these bacteria contributed to the observed growth improvements. Gibberellic acid influences not only plant growth but also factors such as yield, mineral nutrition and nitrogen metabolism (Verma, Ravindran and Kumar, 2016). Moreover, GAs have the capacity to mitigate the impacts of various stresses by enhancing plant tolerance, with effects observed on uptake of ions, distribution and metabolic activities under favourable and unfavourable conditions (Verma, Ravindran and Kumar, 2016). A previous study has shown that reducing endogenous GA levels through the upregulation of GA 2-oxidases (GA2oxs) can enhance drought stress tolerance in rice, resulting in increased plant height (Zhang et al., 2022). Additionally, studies like that of Alonso-Ramírez et al. (2009) on *Arabidopsis thaliana* demonstrated the ability of gibberellic acid to reduce the adverse effects of salinity stress on germination and growth. This hormone's influence extends to its interaction with other plant responses, such as those involving salicylic acid (SA) and osmotic stress. GAs have been applied endogenously to plants to improve responses to osmotic stress, as observed in both *Arabidopsis* and wheat (Ahmad, 2010; Manjili, Sedghi and Pessarakli, 2012).

7.3.3 Cytokinins

Cytokinins (CKs) represent a vital group of plant hormones responsible for maintaining cellular proliferation and differentiation while preventing premature leaf senescence (Schmülling, 2002). These hormones are produced in unripe fruits and seeds and are concentrated in root tips. Their roles encompass stimulating the growth of lateral shoots and participating in cell division (Sosnowski, Truba and Vasileva, 2023). CKs interact with other plant hormones and play crucial roles in various physiological processes. They act as antagonists to ABA and auxin, impacting plant growth and development (Danilova et al., 2016). CKs are instrumental in enabling plant survival under different stress conditions (Danilova et al., 2016). Their involvement in plant responses to salt stress has been observed, with seed priming using CKs to enhance salt tolerance, ultimately supporting normal growth and development through interactions with other hormones such as auxin (Iqbal, Ashraf and Jamil, 2006). Moreover, CKs are suspected to influence photosynthesis processes in plants, aiding in the response to biotic and abiotic stress conditions (Hudeček et al., 2023).

Research conducted by Rivero et al. (2009, 2010) explored the role of CKs in drought situations, revealing increased drought tolerance in transgenic tobacco plants due to elevated CK levels. Additionally, overexpressing CK in transgenic cassava plants improved drought tolerance when compared to their wild-type counterparts (Zhang et al., 2010).

7.3.4 Ethylene

Ethylene (ET) stands as a vital phytohormone deeply entwined in both plant growth and stress responses. It operates as a gaseous hormone, significantly influencing the physical structure and essential physiological processes that underpin a plant's survival. ET takes part in various pivotal processes such as seed germination, flower development and fruit ripening. Its importance further extends to orchestrating plant responses to an array of environmental stressors, including cold stress and pathogen attacks. Additionally, ET plays a substantial role in the plant's reaction to nutrient deficiencies and physical injuries (Abeles, Morgan and Saltveit, 1992; Kendrick and Chang, 2008). Research conducted by Hahn et al. (2008) shed light on the significance of the root cap in regulating elongation growth and root hair formation induced by ET. The absence of inhibition in elongation growth when ET was externally applied to maize plants after removing the root cap underscored the root cap's role. ET also engages in intricate interactions with ABA, particularly in the context of primary root growth. ABA acts upon cortical cells in the elongation zone to inhibit growth, and this process is dependent on the ET signalling pathway (Dietrich et al., 2017). Together, ET and ABA are pivotal in plant responses to environmental stress, with ABA biosynthesis taking place in the roots and subsequently being transported to the leaves. This redistribution leads to higher ABA concentrations in the leaves during stressful conditions, resulting in stomatal closure, reduced transpiration rates and, ultimately, reduced growth (Peleg and Blumwald, 2011).

7.3.5 Abscisic Acid

ABA, a plant-growth regulator, has a rich history dating back to its discovery in 1963 during investigations into cotton fruit abscission. Over time, ABA has revealed its multifaceted role in plant development, impacting seed and shoot dormancy, germination and various morphogenetic processes. It holds the distinction of being the first gaseous plant hormone to be identified (Finkelstein, 2013). ABA's influence is particularly pronounced during the plant's response to environmental stressors, when it enhances stress responses and adaptation. Under stressful conditions, ABA modulates responsive genes, amplifying tolerance responses. However, an excessive buildup of ABA during stress can result in reduced growth and decreased stress tolerance (Asghar et al., 2019). The application of exogenous ABA has proven effective in improving stress tolerance in several plant species, including *Solanum tuberosum* (Mora-Herrera and Lopez-Delgado, 2007) and wheat (Bano, Ullah and Nosheen, 2012). These applications have been associated with improved antioxidant systems and relative water content. Furthermore, ABA plays a pivotal role during salinity stress, with its concentration increasing in response to salinity in plants such as the tomato (Amjad et al., 2014). This regulation of ABA activity is known to impact ET biosynthesis and signalling, and ET insensitive2 (EIN2), a nuclear protein, acts as a mediator between ET and ABA. Changes in EIN2 expression during salinity stress can influence ABA production and its subsequent effects on plant physiology (Wang et al., 2007). Additionally, the application of ABA to citrus plants has been observed to reduce ET concentration and leaf abscission (Gómez-Cadenas et al., 2002).

7.4 PHYTOCHEMICALS IN PLANT DEFENCE

In the ongoing battle against ever-evolving plant pest and environmental stressors, plants have evolved a sophisticated defence arsenal to ensure their survival and productivity (War et al., 2012). One of the key players in this intricate defence system is phytohormones, which play a crucial role in coordinating responses to external threats, such as herbivore attacks, microbial infections and adverse environmental conditions (Zhao et al., 2021; Vaishnav and Chowdhury, 2023). Understanding the role of phytohormones in plant defence has opened up new avenues for developing sustainable and eco-friendly strategies to enhance crop protection and resilience (Chen and Pang, 2023). In this section, we will explore the fascinating world of phytohormones and their involvement in orchestrating plant defence mechanisms. We will delve into the way in which phytohormones interact to activate defence responses and examine how modern agricultural practices leverage this knowledge to bolster plant immunity, reduce the reliance on chemical pesticides and promote sustainable crop management. As we unravel the mysteries behind phytohormone-mediated plant defence strategies, we gain valuable insights into harnessing plants' innate power to thrive in the face of adversity.

7.4.1 Jasmonic Acid

Jasmonic acid (JA) and its derivatives, collectively referred to as jasmonates, are integral plant hormones that play a pivotal role in fortifying plant resistance against herbivores (Howe and Jander, 2008). These compounds, encompassing methyl jasmonate (MeJA), cis-jasmone (CJ) and prohydrojasmon (PDJ), have demonstrated their ability to induce and prime defence responses against insect attacks (Birkett et al., 2000; Thaler et al., 2001; Uefune, Ozawa and Takabayashi, 2014). Upon application, jasmonates activate genes associated with defence mechanisms and secondary metabolites, culminating in enhanced plant defence (Vijayan et al., 1998; Wasternack, 2014). Furthermore, they modulate volatile emissions, attracting natural enemies that prey on pests (Rodriguez-Saona et al., 2001). Additionally, jasmonates stimulate the production of extrafloral nectar, reinforcing the "attract and reward strategy" between plants and their natural enemies (Rodriguez-Saona, Blaauw and Isaacs, 2012). The application of JA and its derivatives holds significant promise in enhancing plant defences and promoting sustainable pest management practices in agriculture (Ali et al., 2021; Ali, Bayram, et al., 2023). The efficacy of jasmonate treatment depends on factors such as timing, tested cultivars and application dosage (Thaler, 1999; Thaler et al., 2002; Doostkam *et al.*, 2023). Additionally, plants treated with JA emit volatile blends akin to those released by herbivore-damaged plants, attracting natural enemies and improving overall pest management (Boland et al., 1995; Dicke et al., 1999; Ament et al., 2004; Bruinsma et al., 2009). These compounds can induce both direct and indirect plant defences through root and shoot applications (Li et al., 2013; Pierre et al., 2013). Furthermore, jasmonate treatment enhances the production of extrafloral nectar, providing essential nourishment for natural enemies and fortifying their fitness (Gols, Posthumus and Dicke, 1999; Gols et al., 2003; Bruinsma et al., 2008). In conclusion, JA and jasmonates play a crucial role in plant defence, with their application offering promising solutions for sustainable pest management and improved agriculture practices.

7.4.2 Methyl Jasmonate

MeJA is a vital component in the arsenal of plant defences against herbivores. It is a volatile compound, specifically a methyl ester, which occurs naturally in plants and plays a pivotal role in coordinating plant responses to stress and attacks by herbivores (Farmer and Ryan, 1990; Creelman and Mullet, 1995; Seo et al., 2001; Wasternack, 2007). MeJA is synthesised through a pathway involving the enzymatic conversion of (JA into its methyl ester form (MeJA). This conversion is an essential step in the plant's defence signalling network (Yao and Tian, 2005; Anjum et al., 2011; Jiang and Yan, 2018). MeJA is a key player in inducing plant defence mechanisms against herbivores. It triggers the activation of defence enzymes, including superoxide dismutase (SOD), phenylalanine ammonia lyase (PAL) and polyphenol oxidase (PPO). Additionally, MeJA induces the production of protease inhibitors such as trypsin inhibitor and chymotrypsin inhibitor, which disrupt herbivores' feeding and digestion (Lomate and Hivrale, 2012). MeJA also stimulates the production of defence-related compounds in MeJA-treated plants, including flavonoids, glucosinolates and a range of phenolic compounds (Belhadj et al., 2006; Bi et al., 2007). These compounds act as chemical deterrents, making the plant less palatable to herbivores and reducing their overall performance (Baldwin, 1996; Wei et al., 2021).

MeJA enhances plant defences both directly and indirectly. In direct defence, it affects herbivores' performance, reducing their reproduction, survival, settlement, feeding and physiological functions (Erbilgin et al., 2006; Heijari et al., 2008; Zhao et al., 2011; Bayram and Tonğa, 2018; Stella de Freitas, Stout and Sant'Ana, 2019; Tonga et al., 2022). Indirectly, MeJA increases the emission of plant volatiles, which serve as signals to attract natural enemies and predators of herbivores, enhancing overall pest control (Rodriguez-Saona et al., 2001; Martin, Gershenzon and Bohlmann, 2003; Erbilgin et al., 2006; Hare, 2007; Schiebe et al., 2012; Semiz et al., 2012; Rodriguez-Saona, Polashock and Malo, 2013; Bayram and Tonga, 2018; Tonga et al., 2022). MeJA can be applied both exogenously and endogenously. Its exogenous application, as a plant defence elicitor, shows promise for pest management and sustainable agriculture practices. The manipulation of MeJA signalling pathways provides a valuable tool for enhancing plant resistance to herbivores and promoting eco-friendly pest control strategies in agriculture.

7.4.3 Cis-Jasmone

CJ is a naturally occurring stress signal primarily induced by herbivory or external stimuli (Koch, Bandemer and Boland, 1997; Tanaka et al., 2009). It activates unique genes distinct from the MeJA-induced signalling pathway (Matthes et al., 2010) and induces plant defence against pests, both through direct and indirect mechanisms (Blassioli Moraes et al., 2008;

Bruce et al., 2008; Ali et al., 2021). CJ alters plant volatile emissions, making them less appealing to herbivores while attracting natural enemies, thus reducing herbivore performance and population density (Bruce, Pickett and Smart, 2003; Pickett et al., 2007b; Bruce et al., 2008; Dewhirst et al., 2012; Tonǧa et al.. 2020; Ali et al., 2021). Moreover, CJ enhances tritrophic interactions by attracting natural enemies and primes plants for the increased production of defensive volatile organic compounds, such as (E)-2-hexenal, 6-methyl-5-hepten-2-one (MHO), (Z)-3-hexenyl acetate, myrcene, (E)-ocimene, (E)-4,8-dimethyl-1,3,7-nonatriene (DMNT), methyl salicylate (MeSA), caryophyllene, (E)-β-farnesene and (E,E)-4,8, 12-trimethyl-1,3,7,11-tridecatetraene (TMTT) (Hegde et al., 2012; Sobhy et al., 2017; Bayram and Tonǧa, 2018; Tonǧa et al., 2020; Ali et al., 2021). The exogenous application of CJ demonstrates its potential for pest control management and sustainable agriculture, making it a valuable tool in enhancing plant defences against herbivores (Ali, 2023; Ali et al., 2021; Sobhy, Lou and Bruce, 2022).

7.4.4 Salicylic Acid

SA, a natural phytohormone, is a pivotal component in a plant's immune system, primarily serving as a defence mechanism against biotrophic pathogens (Raskin, 1992; Loake and Grant, 2007; An and Mou, 2011; Ding and Ding, 2020). When a plant confronts pathogenic threats, SA rapidly becomes an essential element of its defence strategy (Lefevere, Bauters and Gheysen, 2020). This molecule is synthesise through the phenylpropanoid pathway and undergoes a rapid increase in response to pathogen invasion (Dempsey, Shah and Klessig, 1999; Hayat, Ali and Ahmad, 2007). This surge in SA levels activates a wide array of defence-related genes and metabolic pathways, triggering a cascade of defence mechanisms aimed at safeguarding the plant from pathogenic infections (Tripathi, Raikhy and Kumar, 2019; Mohamed, El-Shazly and Badr, 2020). SA derivatives, such as MeSA, contribute to plant defence by participating in the synthesis of defensive compounds and activating defence-related genes (Park et al., 2007).

For decades, SA has been recognised as a crucial component in the establishment of systemic acquired resistance (SAR), a comprehensive defence mechanism that bestows long-lasting immunity against a wide range of pathogens (Malamy et al., 1990; Gaffney et al., 1993). Upon pathogen recognition, SA levels increase locally at the infection site and subsequently trigger systemic signalling, leading to heightened resistance

throughout the entire plant (Zhu et al., 2014; Liang et al., 2022). This signalling cascade involves the activation of pathogenesis-related (PR) genes,
which encode antimicrobial proteins and other defence-related genes that
fortify the plant's immune system (Wang et al., 2012; Zhu et al., 2014; Ding
et al., 2022). SA-mediated defence is particularly effective against biotrophic pathogens, which rely on living host cells for their nutrition and survival (An and Mou, 2011). Furthermore, SA's role extends beyond defence
against pathogens. It also contributes to mitigating abiotic stresses, including drought, heat and heavy metal toxicity (Khan et al., 2015). SA signalling pathways interact with other hormonal pathways, such as ABA and
ET, to fine-tune the plant's response to various stressors (Yang et al., 2015;
Nguyen et al., 2016). These attributes make SA and its derivatives invaluable in the context of pest control and sustainable agriculture applications.
SA derivatives, such as MeSA and acetyl SA (aspirin), also contribute to
plant defence responses. MeSA is a volatile compound released by plants
upon pathogen attack (Malamy et al., 1990; Gaffney et al., 1993). It serves
as a mobile signal that not only induces defence in neighbouring plants
but also attracts beneficial insects, contributing to natural pest control
(Park et al., 2007; Singewar, Fladung and Robischon, 2021; Ali, Wei, et al.,
2023). Aspirin, derived from SA, has been found to enhance plant defence
mechanisms and increase resistance to various pathogens (White, 1979;
Senaratna et al., 2000).

7.4.5 Brassinosteroids

Brassinosteroids (BRs), a class of steroid hormones, in regulating diverse
physiological processes in plants, including growth, development, reproduction and stress responses (Krishna, 2003; Hayat et al., 2019; Nolan
et al., 2020). Notably, BRs have been demonstrated to play a vital role in
enhancing plant defences against abiotic stress (Krishna, 2003; Nolan et
al., 2020). The role of BRs extends beyond abiotic stress responses, as these
steroidal plant hormones also play a significant role in regulating biotic
stress responses in plants (Vardhini et al., 2010). Biotic stress encompasses
challenges posed by various pests and pathogens that can severely impact
plant health and crop productivity (Gull, Lone and Wani, 2019). BRs have
been found to be instrumental in orchestrating plant defence mechanisms
against these biotic stressors (Khripach, Zhabinskii and de Groot, 1998;
Campos and Peres, 2012; Ali, 2017; Hussain et al., 2020). Upon recognising pathogen-associated molecular patterns (PAMPs), plants activate
front-line defence mechanisms known as PAMP-triggered immunity

(PTI) (Zipfel, 2009; Zipfel and Robatzek, 2010). BRs have been shown to enhance PTI responses by promoting the accumulation of reactive oxygen species (ROS) (Xia et al., 2009), callose deposition (Xiong et al., 2020; Benitez-Alfonso and Caño-Delgado, 2023) and the expression of defence-related genes (Xia et al., 2011; Xiong et al., 2020). Moreover, BRs are implicated in the regulation of effector-triggered immunity (ETI), a more sophisticated defence mechanism activated upon recognition of pathogen effectors by specific plant resistance (R) proteins (Naveed et al., 2020). By enhancing the recognition and signalling cascades initiated by R proteins, BRs contribute to the hypersensitive response (HR) and the establishment of SAR, ensuring plants can mount effective immune responses tailored to the specific pathogens they encounter (He et al., 2000; Grant and Lamb, 2006; Yu, Zhao and He, 2018). Additionally, the interplay between BRs and other phytohormones, such as SA and JA, further enhances the complexity of plant immune responses, fine-tuning the balance between SA- and JA-dependent defence strategies to optimise plant defence against different types of biotic stressors (Divi, Rahman and Krishna, 2010; Yang et al., 2019). Thus, BRs emerge as key regulators in the intricate network of plant immune responses, offering promising opportunities for developing eco-friendly and targeted strategies to safeguard crops, enhance resistance and ensure agricultural sustainability in the face of evolving pests and pathogens (Krishna, 2003; Manghwar et al., 2022).

7.4.6 Systemin and Defensins

Systemin and defensins are other important key players that play a vital role in activating innate immunity in plants and orchestrating defence mechanisms against herbivores and microbial invaders, respectively (Stotz, Thomson and Wang, 2009; Choi and Klessig, 2016). Systemin, a small peptide, and defensins, antimicrobial peptides, trigger a cascade of events that lead to the synthesis of JA and the activation of defence-related genes, providing rapid and targeted protection against pests and pathogens (Ryan, 2000; Choi and Klessig, 2016). The structural diversity of defensins allows them to combat a wide range of pathogens, and their crosstalk with other defence pathways fine-tunes the plant's response to specific threats (Odintsova et al., 2019). Understanding the molecular mechanisms of systemin and defensins opens avenues for sustainable pest and pathogen management in agriculture, including the development of biopesticides and elicitor-based strategies (Anderson et al., 2016; Vincent, Rafiqi and Job, 2020; Leannec-Rialland et al., 2022). Harnessing these natural defence inducers offers promising eco-friendly and targeted solutions to ensure crop health, resistance and global food production sustainability amidst changing environmental challenges.

7.5 CHEMICAL ECOLOGY FOR SUSTAINABLE PEST CONTROL AND DIVERSIFIED AGROECOSYSTEMS

Chemical ecology offers profound insights into the origins, functions and significance of natural chemicals governing communication in both aquatic and terrestrial environments. These relationships are adaptively crucial and have become the focal point of a robust interdisciplinary collaboration between chemists and biologists. This alliance holds great promise in advancing our understanding of ecological interactions, particularly in the context of sustainable pest control in agriculture (Pickett, Wadhams and Woodcock, 1997; Khan et al., 2016; Mbaluto et al., 2020). Semiochemicals, serving as signalling compounds, play a pivotal role in the paradigm shift toward environmentally conscious pest management (Bruce, 2010; Komala, Manda and Seram, 2021). These substances are crucial in various aspects, including pest attraction and elimination, mating disruption, interference with pheromones during mating, use of aggregation pheromones in trap-out procedures and the application of alarm pheromones to deter pests (Suckling et al., 2000; Weinzierl et al., 2005; Ioriatti et al., 2011). The integration of stimulo-deterrent diversionary strategies (SDDS) marks a significant leap, in which semiochemicals, such as sex pheromones and host-masking agents, are strategically utilised to protect harvestable crops while concurrently influencing pest behaviour in trap crops (Cook et al., 2007; Sarkar et al., 2018). This exploration navigates through the multifaceted landscape of semiochemical strategies, encompassing sex pheromones, alarm pheromones and SDDS as indispensable components for a sustainable and effective approach to pest management in agriculture (Miller and Cowles, 1990; Duraimurugan and Regupathy, 2005).

7.5.1 Push-Pull Approach

The push-pull approach is an innovative strategy employed in integrated pest management programs, offering a holistic solution to pest control (Cook, Khan and Pickett, 2007; Khan et al., 2012, 2016). At its core, this strategy utilises a combination of repellent and attractive stimuli to manipulate the distribution and abundance of insect pests and their natural enemies. In practice, the push-pull strategy operates on two principal mechanisms: "push" and "pull." First, pests are repelled or deterred away from the main crop using stimuli that mask host apparency or possess repellent properties (push). Simultaneously, pests are attracted to alternative areas, such as traps or trap crops, using highly apparent and attractive stimuli (pull) (Figure 7.1). For instance, in the management of cereal stemborers in Africa, the push-pull strategy involves trapping stemborers on highly attractant trap plants (pull), while concurrently driving them away from the main crop using repellent intercrops (push). Key components include the use of Napier grass and Sudan grass as trap plants, which attract stemborers, and the incorporation of molasses grass and desmodium as repellent intercrops to deter pest infestation. This strategy has demonstrated significant success in enhancing crop protection while minimising reliance on chemical pesticides. By harnessing the natural interactions between plants and insects, the push-pull approach offers a sustainable and environmentally friendly solution to pest management challenges.

7.5.2 Antifeedants and Alarm Pheromones

Antifeedant compounds, derived from plants, have shown promise against aphid pests and coleopterous pests in arable agriculture. For instance, polygodial, an antifeedant from *Polygonum hydropiper*, demonstrated a more than 70% reduction in the transmission of potato virus Y by the peach-potato aphid (Pickett, 1989). In field trials, applying polygodial on cereals resulted in improved yields comparable to the synthetic pyrethroid insecticide cypermethrin (Prota, Bouwmeester and

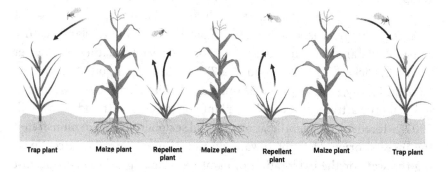

FIGURE 7.1   Schematic representation of the push-pull approach. The strategy employs repellent intercrops (push) to deter pests from the main crop, while simultaneously utilising attractive trap plants (pull) to concentrate pests away from the main crop, facilitating their control.

Jongsma, 2014). Additionally, the integration of the aphid alarm phero-mone (E)-β-farnesene, released when aphids are attacked, along with other population-reducing agents, showed substantial control improve-ments in both glasshouse and field trials against aphid pests (Cook, Khan and Pickett, 2007). For coleopterous pests, antifeedants such as ajugarins in the clerodane class exhibited high effectiveness, particularly against the Chrysomelidae family (Caballero et al., 2001; Sosa and Tonn, 2008). Electrostatic spraying of ajugarins protected mustard plants in simulated field trials, necessitating a combination with the insect growth regulant teflubenzuron for comprehensive pest control (Akhtar, 2003; Swami et al., 2022). Current efforts include developing a new range of antifeedants, the lupulones from the hop plant, *Humulus lupulus*, showcasing strong antifeedant activity against various pests, including mites, with potential applications in protecting crops (De Acutis, 2019).

7.5.3 Stimulo-Deterrent Diversionary Strategies

The quest for innovative and sustainable approaches to pest control has led to the development of SDDS. The identification of insect sex phero-mones sparked interest in exploring the potential for long-range selection of host plants through plant-derived semiochemicals (Miller and Cowles, 1990; Ben-Issa, Gomez and Gautier, 2017; Kamala Jayanthi, Raghava and Kempraj, 2020). The primary rhinaria on the fifth and sixth segments of the aphid antenna play a crucial role in detecting plant chemicals, and coupled gas chromatography-single cell recordings (GC-SCR), enabled the identification of host plant attractants. Notably, certain cells responded to compounds from non-host plants, acting as repellents for specific aphid species. Investigations into the black bean aphid, *Aphis fabae*, revealed its ability to detect isothiocyanates, typical of Cruciferae plants, acting as repellents and masking attractancy of host plant volatiles (Pickett et al., 1992; Pickett and Glinwood, 2007). Similarly, compounds from the Labiatae family, such as (-)-(1*R*,5*S*)-myrtenal, were identified as repellents, reducing the attractiveness of host plant volatiles (Pickett and Glinwood, 2007). Moreover, the identification of aphid sex pheromones demonstrated the insects' oriented flight toward distant semiochemical sources, pav-ing the way for the exploration of plant-derived semiochemicals in host plant selection. In the case of the bird-cherry aphid (*Rhopalosiphum padi*), which colonises *Prunus padus* as its primary host and migrates to cereal crops in the spring, MeSA emerged as a promising candidate for repellent activity. Field trials on barley in Sweden showed a 50% reduction in *R.*

padi populations using MeSA, released through emulsifiable concentrate or slow-release vials. Further trials on spring-sown wheat demonstrated consistent reductions in cereal aphid populations, indicating the potential of MeSA as an effective semiochemical for aphid pest control. The evolving understanding of SDDS opens new avenues for sustainable and environmentally friendly pest management strategies (Pickett, Wadhams and Woodcock, 1997; Pickett and Glinwood, 2007).

7.5.4 Harnessing Priming for Sustainable Agriculture

Plant defence priming is akin to a green vaccination strategy. It involves intentionally exposing plants to mild stress or defence elicitors inducing an alarmed state known as the primed state. Despite lacking memory cells, plants exhibit heightened sensitivity and responsiveness to subsequent stimuli, enabling faster and stronger defence responses (Ton et al., 2007; Conrath et al., 2015). The priming mechanism involves the perception of stimuli by the plant, triggering physiological, molecular and epigenetic changes. This results in enhanced readiness to combat future challenges, marked by the accumulation of various substances and the activation of defence-related genes. The maintenance of defence priming in plants is a long-term phenomenon involving epigenetic modifications, enabling plants to build a memory of challenges and modulating gene responsiveness (Ton et al., 2007; Hilker and Meiners, 2010). The durability of priming within a generation is notable, with primed effects persisting even after removing the priming agent. This long-lasting impact has been observed in various plant species, such as soybean and barley, showcasing the resilience of priming effects. Moreover, priming exhibits transgenerational immune priming, in which stressed parents give birth to progeny with enhanced defence responses (Cooper and Ton, 2022). This inheritance of priming underscores the role of environmental stressors in inducing epigenetic changes and ensuring augmented defence responses in subsequent plant generations. While defence priming provides robust protection against various stresses, it is not without costs. Resource allocation, autotoxicity, ecological impacts and growth penalties are among the challenges. Studies suggest that the benefits of priming often outweigh the costs, especially in the presence of disease pressure (Gupta and Bar, 2020; Bagheri and Fathipour, 2021). The innovative and sustainable approach of utilising priming inducing factors represents a promising trajectory for enhancing crop protection, offering a paradigm shift toward a resilient and sustainable agriculture landscape (Christou, Agathokleous and

Fotopoulos, 2022; Salam et al., 2022). In conclusion, integrating plant defence priming into agricultural practices emerges as a strategic cornerstone for achieving sustainable agriculture. The innate immune capacity of plants, when harnessed through priming, provides an effective, economical and environmentally friendly means for crop improvement, ensuring global food security while navigating the challenges of a changing climate and growing population.

7.5.6 Pollination

Chemical ecology plays a pivotal role in enhancing crop pollination by utilising floral volatiles to attract and recruit pollinators. The emission of specific chemical compounds from flowers serves as a potent signal that lures essential pollinators, such as bees, as well as biocontrol predators and parasitoids (Kumawat et al., 2021; Colazza, Peri and Cusumano, 2023). This intricate interplay of chemical communication fosters a mutually beneficial relationship, ensuring effective and sustainable pollination in agricultural settings. Floral volatiles, encompassing more than 1,700 known compounds, serve as crucial players in plant–plant communication and have a profound impact on pollination dynamics in diversified agroecosystems (Dötterl and Vereecken, 2010; Bohman et al., 2016).

7.5.7 Natural Enemies

Chemical ecology plays a pivotal role in bolstering the recruitment of natural enemies, such as predators and parasitoids, in agricultural ecosystems. Floral volatiles, a key component, are volatile organic compounds emitted by plants, particularly during their flowering stages. These compounds act as chemical signals, attracting beneficial organisms to the vicinity of the plant (Pichersky and Gershenzon, 2002; Colazza, Peri and Cusumano, 2023). Herbivore-induced plant volatiles (HIPVs) represent a specific subset of plant volatiles released in response to herbivore attacks. When plants experience herbivore feeding, they release HIPVs as part of their defensive strategy. These volatile compounds serve as an alarm system, signalling nearby predators and parasitoids to the presence of herbivores. This mechanism enhances the efficiency of natural enemies in locating and controlling herbivorous pests (Khan et al., 2008; Peñaflor and Bento, 2013). Furthermore, the utilisation of plant-derived chemicals as defence elicitors is a noteworthy aspect of chemical ecology (Garcia-Brugger et al., 2006). These chemicals, often produced by plants in response to stress or herbivore damage, induce a cascade of biochemical reactions within the plant, resulting in the synthesis of various defensive compounds. This induced chemical defence mechanism not only fortifies the plant against further herbivore attacks but also contributes to the emission of additional volatile organic compounds (Ali et al., 2021; Meena et al., 2022; Ali, Wei, et al., 2023). These emitted volatiles act as attractants, drawing in natural enemies of herbivorous insects and fostering a more robust and ecologically balanced system. In summary, the intricate interplay of floral volatiles, HIPVs and plant-derived chemicals as defence elicitors highlights the nuanced ways in which chemical ecology supports the recruitment of biocontrol agents, ultimately contributing to sustainable and effective pest management in agriculture.

7.5.7 Traps and Lures in Agriculture

Identification of chemical compounds and behavioural responses of insects to these compounds lead to the foundation of developing traps and lures for pest management approaches. Development of traps and lures is one the practical applications in which chemical ecology plays as a vital role in agriculture pest management (Weinzierl et al., 2005; Spears et al., 2016). The primary reason is that it helps in monitoring and predicting pest populations, which is crucial for implementing appropriate control measures (Mamay and Mutlu, 2019). Traps and lures use chemical cues, such as pheromones, to attract pest insects and trap them, making it easy to determine their presence and density (Tan et al., 2014). Additionally, using traps and lures in agriculture reduces the need for indiscriminate pesticide applications, which may harm beneficial insects and the environment. By targeting specific pests, the use of pesticides can be limited, reducing the risk of resistance development (Sunamura et al., 2011). Furthermore, traps and lures are cost-effective and easy to use. They require little maintenance, are easy to deploy and can be used to trap large numbers of pests over an extended period. Farmers in developing countries, where the cost of insecticides and other pest management tools is a significant challenge, can benefit greatly from such affordable and accessible technologies (El-Sayed et al., 2009; Norbury et al., 2014). In conclusion, the development of traps and lures in agriculture pest management from the perspective of chemical ecology is essential for sustainable agriculture, reduced pesticide use and cost-effectiveness. Therefore, policymakers, researchers and farmers should invest in the development and utilisation of traps and lures to achieve sustainable and environmentally friendly pest management practices.

7.6 CONCLUSION

In conclusion, the exploration of chemical ecology within the context of agriculture sheds light on the intricate interactions between plants, insects, fungi and bacteria. Phytochemicals play a central role in these relationships, serving as multifunctional tools that plants employ to communicate, defend and adapt in their environment. The study of plant chemical defences against insects highlights the nuanced strategies that plants have evolved to deter herbivores and shape the dynamics of pest populations. Similarly, the examination of chemical defences against fungal and bacterial attacks unveils the diverse mechanisms through which plants fortify themselves against microbial threats, emphasising the complex chemical dialogues occurring within the rhizosphere. Phytotoxins emerge as key players in plant pathology, illustrating the dual role of chemical compounds as both defences and weapons in the intricate battlefield of plant–microbe interactions. The chemical basis of biological control unveils the potential of harnessing nature's own chemical cues to manage pest populations, showcasing the promise of biopesticides and sustainable pest management strategies. The innovative push-pull approach, rooted in chemical ecology principles, underscores the power of manipulating insect behaviours through attractants and repellents to create balance in agricultural ecosystems. Furthermore, the integration of trap and lure systems underscores the practical applications of chemical ecology in pest management, providing an effective means to monitor and mitigate pest infestations. These tools leverage the knowledge of insect semiochemicals to lure pests away from crops while guiding them into traps, thereby contributing to reduced chemical inputs and increased crop yields. In essence, this chapter delves into the intricate web of chemical interactions that govern the relationships between plants, insects, microbes and the environment in the realm of agriculture. By unravelling the chemical language underlying these interactions, we gain insights that not only enhance our understanding of natural systems but also pave the way for innovative and sustainable approaches to agricultural practices. The integration of chemical ecology principles into agriculture has the potential to revolutionise pest management, reduce environmental impacts and contribute to a more harmonious coexistence between humans, crops and the natural world.

REFERENCES

Abd El-Ghany, N. M. (2020) 'Pheromones and chemical communication in insects', *Pests, weeds and diseases in agricultural crop and animal husbandry production*. IntechOpen, pp. 16–30.

Abeles, F. B., Morgan, P. W. and Saltveit, M. E. (1992) 'Regulation of ethylene production by internal, environmental, and stress factors', *Ethylene in plant biology*, pp. 56–119.

De Acutis, L. (2019) 'Characterisation of different hop ecotypes (Humulus lupulus, L.) in Central Italy and evaluation of the biological activity of their extracts, EO and components against Sitophilus granarius (L.)'.

Ahmad, I. *et al.* (2019) 'Effects of plant growth regulators on seed filling, endogenous hormone contents and maize production in semiarid regions', *Journal of Plant Growth Regulation*, 38, pp. 1467–1480.

Ahmad, P. (2010) 'Growth and antioxidant responses in mustard (Brassica juncea L.) plants subjected to combined effect of gibberellic acid and salinity', *Archives of Agronomy and Soil Science*, 56(5), pp. 575–588.

Ahmed, A. and Hasnain, S. (2014) 'Auxins as one of the factors of plant growth improvement by plant growth promoting rhizobacteria', *Polish Journal of Microbiology*, 63(3), p. 261.

Akhtar, Y. (2003) 'The effects of feeding experience with antifeedants on larval feeding and adult oviposition behavior in generalist and specialist herbivores'. University of British Columbia.

Ali, B. (2017) 'Practical applications of brassinosteroids in horticulture—some field perspectives', *Scientia Horticulturae*, 225, pp. 15–21.

Ali, J. (2023) 'The peach potato Aphid (Myzus persicae): ecology and management', 1, p. 132. doi: 10.1201/9781003400974.

Ali, J. *et al.* (2021) 'Effects of cis-Jasmone treatment of brassicas on interactions with Myzus persicae Aphids and their parasitoid diaeretiella rapae', *Frontiers in Plant Science*, 12. doi: 10.3389/fpls.2021.711896.

Ali, J., Bayram, A., *et al.* (2023) 'Peach–potato Aphid Myzus persicae: current management strategies, challenges, and proposed solutions', *Sustainability*, 15(14), p. 11150.

Ali, J., Wei, D., *et al.* (2023) 'Exogenous application of methyl salicylate induces defence in brassica against peach potato Aphid Myzus persicae', *Plants*, 12(9), p. 1770.

Alonso-Ramírez, A. *et al.* (2009) 'Evidence for a role of gibberellins in salicylic acid-modulated early plant responses to abiotic stress in Arabidopsis seeds', *Plant Physiology*, 150(3), pp. 1335–1344.

Ament, K. *et al.* (2004) 'Jasmonic acid is a key regulator of spider mite-induced volatile terpenoid and methyl salicylate emission in tomato', *Plant Physiology*, 135(4), pp. 2025–2037. doi: 10.1104/pp.104.048694.

Amjad, M. *et al.* (2014) 'Integrating role of ethylene and ABA in tomato plants adaptation to salt stress', *Scientia Horticulturae*, 172, pp. 109–116.

An, C. and Mou, Z. (2011) 'Salicylic acid and its function in plant immunity F', *Journal of Integrative Plant Biology*, 53(6), pp. 412–428.

Anderson, J. A. *et al.* (2016) 'Emerging agricultural biotechnologies for sustainable agriculture and food security', *Journal of Agricultural and Food Chemistry*, 64(2), pp. 383–393.

Andolfi, A. *et al.* (2011) 'Phytotoxins produced by fungi associated with grapevine trunk diseases', *Toxins*, 3(12), pp. 1569–1605.

Anjum, S. A. *et al.* (2011) 'Methyl jasmonate-induced alteration in lipid peroxidation, antioxidative defence system and yield in soybean under drought', *Journal of Agronomy and Crop Science*, 197(4), pp. 296–301.

Asghar, M. A. *et al.* (2019) 'Crosstalk between abscisic acid and auxin under osmotic stress', *Agronomy Journal*, 111(5), pp. 2157–2162.

Bagheri, A. and Fathipour, Y. (2021) 'Induced resistance and defense primings', *Molecular approaches for sustainable insect pest management*, pp. 73–139.

Baldwin, I. T. (1996) 'Methyl jasmonate-induced nicotine production in Nicotiana attenuata: inducing defenses in the field without wounding', *Entomologia Experimentalis et Applicata*, 80(1), pp. 213–220. doi: 10.1111/j.1570-7458.1996.tb00921.x.

Bano, A., Ullah, F. and Nosheen, A. (2012) 'Role of abscisic acid and drought stress on the activities of antioxidant enzymes in wheat'. *Plant, Soil and Environment*, 58(4), pp. 181–185.

Bayram, A. and Tonğa, A. (2018) 'cis-Jasmone treatments affect pests and beneficial insects of wheat (Triticum aestivum L.): the influence of doses and plant growth stages', *Crop Protection*, 105, pp. 70–79.

Belhadj, A. *et al.* (2006) 'Methyl jasmonate induces defense responses in grapevine and triggers protection against Erysiphe necator', *Journal of Agricultural and Food Chemistry*, 54(24), pp. 9119–9125.

Ben-Issa, R., Gomez, L. and Gautier, H. (2017) 'Companion plants for aphid pest management', *Insects*, 8(4), p. 112.

Benitez-Alfonso, Y. and Caño-Delgado, A. I. (2023) 'Brassinosteroids en route', *Nature Chemical Biology*, pp. 1–2.

Bi, H. H. *et al.* (2007) 'Rice allelopathy induced by methyl jasmonate and methyl salicylate', *Journal of Chemical Ecology*, 33(5), pp. 1089–1103.

Birkett, M. A. *et al.* (2000) 'New roles for cis-jasmone as an insect semiochemical and in plant defense', *Proceedings of the National Academy of Sciences of the United States of America*, 97(16), pp. 9329–9334. doi: 10.1073/pnas.160241697.

Blassioli Moraes, M. C. *et al.* (2008) 'cis-Jasmone induces accumulation of defence compounds in wheat, Triticum aestivum', *Phytochemistry*, 69(1), pp. 9–17. doi: 10.1016/j.phytochem.2007.06.020.

Bohman, B. *et al.* (2016) 'Pollination by sexual deception—it takes chemistry to work', *Current Opinion in Plant Biology*, 32, pp. 37–46.

Boland, W. *et al.* (1995) 'Jasmonic acid and coronatin induce odor production in plants', *Angewandte Chemie International Edition in English*, 34(15), pp. 1600–1602. doi: 10.1002/anie.199516001.

Bos, J. I. B. *et al.* (2010) 'A functional genomics approach identifies candidate effectors from the aphid species Myzus persicae (green peach aphid)', *PLoS Genetics*, 6(11). doi: 10.1371/journal.pgen.1001216.

Bruce, T. J. A. *et al.* (2008) 'cis-Jasmone induces Arabidopsis genes that affect the chemical ecology of multitrophic interactions with aphids and their parasitoids', *Proceedings of the National Academy of Sciences of the United States of America*, 105(12), pp. 4553–4558. doi: 10.1073/pnas.0710305105.

Bruce, T. J. A. (2010) 'Exploiting plant signals in sustainable agriculture', pp. 215–227. doi: 10.1007/978-3-642-12162-3_12.

Bruce, T. J. A. and Pickett, J. A. (2011) 'Perception of plant volatile blends by herbivorous insects - Finding the right mix', *Phytochemistry*, 72(13), pp. 1605–1611. doi: 10.1016/j.phytochem.2011.04.011.

Bruce, T., Pickett, J. and Smart, L. (2003) 'cis-Jasmone switches on plant defence against insects', *Pesticide Outlook*, 14(3), pp. 96–98.

Bruinsma, M. *et al.* (2008) 'Differential effects of jasmonic acid treatment of Brassica nigra on the attraction of pollinators, parasitoids, and butterflies', *Entomologia Experimentalis Et Applicata*, 128(1), pp. 109–116. doi: 10.1111/j.1570-7458.2008.00695.x.

Bruinsma, M. *et al.* (2009) 'Jasmonic acid-induced volatiles of Brassica oleracea attract parasitoids: effects of time and dose, and comparison with induction by herbivores', *Journal of Experimental Botany*, 60(9), pp. 2575–2587. doi: 10.1093/jxb/erp101.

Brzozowski, L. J. *et al.* (2020) 'Attack and aggregation of a major squash pest: parsing the role of plant chemistry and beetle pheromones across spatial scales', *Journal of Applied Ecology*, 57(8), pp. 1442–1451.

Caballero, C. *et al.* (2001) 'Effects of ajugarins and related neoclerodane diterpenoids on feeding behaviour of Leptinotarsa decemlineata and Spodoptera exigua larvae', *Phytochemistry*, 58(2), pp. 249–256.

Cabello-Conejo, M. I., Prieto-Fernández, Á. and Kidd, P. S. (2014) 'Exogenous treatments with phytohormones can improve growth and nickel yield of hyperaccumulating plants', *Science of the Total Environment*, 494, pp. 1–8.

Campos, M. L. and Peres, L. E. P. (2012) 'Brassinosteroids as mediators of plant biotic stress responses', *Brassinosteroids: pratical applications in agriculture and human health*. Bentham Science Publishers, pp. 35–43.

Chen, J. and Pang, X. (2023) 'Phytohormones unlocking their potential role in tolerance of vegetable crops under drought and salinity stresses', *Frontiers in Plant Science*, 14, p. 1121780.

Choe, J. C. (2019) *Encyclopedia of animal behavior*. Academic Press.

Choi, H. W. and Klessig, D. F. (2016) 'DAMPs, MAMPs, and NAMPs in plant innate immunity', *BMC Plant Biology*, 16, pp. 1–10.

Christou, A., Agathokleous, E. and Fotopoulos, V. (2022) 'Safeguarding food security: hormesis-based plant priming to the rescue', *Current Opinion in Environmental Science & Health*, 28, p. 100374.

Cingel, A. *et al.* (2016) 'Extraordinary adaptive plasticity of Colorado potato beetle: "Ten-Striped Spearman" in the era of biotechnological warfare', *International Journal of Molecular Sciences*, 17(9), p. 1538.

Colazza, S., Peri, E. and Cusumano, A. (2023) 'Chemical ecology of floral resources in conservation biological control', *Annual Review of Entomology*, 68, pp. 13–29.

Conrath, U. *et al.* (2015) 'Priming for enhanced defense', *Annual Review of Phytopathology*, 53(1), pp. 97–119. doi: 10.1146/annurev-phyto-080614-120132.

Cook, S. M., Khan, Z. R. and Pickett, J. A. (2007) 'The use of push-pull strategies in integrated pest management', *Annual Review of Entomology*, 52(1), pp. 375–400.

Cooper, A. and Ton, J. (2022) 'Immune priming in plants: from the onset to transgenerational maintenance', *Essays in Biochemistry*, 66(5), pp. 635–646.

Creelman, R. A. and Mullet, J. E. (1995) 'Jasmonic acid distribution and action in plants: regulation during development and response to biotic and abiotic stress', *Proceedings of the National Academy of Sciences of the United States of America*, 92(10), pp. 4114–4119. doi: 10.1073/pnas.92.10.4114.

Danilova, M. N. *et al.* (2016) 'Molecular and physiological responses of Arabidopsis thaliana plants deficient in the genes responsible for ABA and cytokinin reception and metabolism to heat shock', *Russian Journal of Plant Physiology*, 63, pp. 308–318.

Dempsey, D. A., Shah, J. and Klessig, D. F. (1999) 'Salicylic acid and disease resistance in plants', *Critical Reviews in Plant Sciences*, 18(4), pp. 547–575.

Després, L., David, J. P. and Gallet, C. (2007) 'The evolutionary ecology of insect resistance to plant chemicals', *Trends in Ecology and Evolution*, 22(6), pp. 298–307. doi: 10.1016/j.tree.2007.02.010.

Dewhirst, S. Y. *et al.* (2012) 'Activation of defence in sweet pepper, Capsicum annum, by cis-jasmone, and its impact on aphid and aphid parasitoid behaviour', *Pest Management Science*, 68(10), pp. 1419–1429. doi: 10.1002/ps.3326.

Dicke, M. *et al.* (1999) 'Jasmonic acid and herbivory differentially induce carnivore-attracting plant volatiles in Lima bean plants', *Journal of Chemical Ecology*, 25(8), pp. 1907–1922.

Dicke, M. (2000) 'Chemical ecology of host-plant selection by herbivorous arthropods: a multitrophic perspective', *Biochemical Systematics and Ecology*, 28(7), pp. 601–617.

Dietrich, D. *et al.* (2017) 'Root hydrotropism is controlled via a cortex-specific growth mechanism', *Nature Plants*, 3(6), pp. 1–8.

Ding, L.-N. *et al.* (2022) 'Plant disease resistance-related signaling pathways: recent progress and future prospects', *International Journal of Molecular Sciences*, 23(24), p. 16200.

Ding, P. and Ding, Y. (2020) 'Stories of salicylic acid: a plant defense hormone', *Trends in Plant Science*, 25(6), pp. 549–565.

Divi, U. K., Rahman, T. and Krishna, P. (2010) 'Brassinosteroid-mediated stress tolerance in Arabidopsis shows interactions with abscisic acid, ethylene and salicylic acid pathways', *BMC Plant Biology*, 10, pp. 1–14.

Doostkam, M., Sohrabi, F., Modarresi, M., Kohanmoo, M.A. and Bayram, A., (2023). 'Genetic variation of cucumber (*Cucumis sativus* L.) cultivars to exogenously applied jasmonic acid to induce resistance to *Liriomyza sativae*', *Arthropod-Plant Interactions*, 17(3), pp. 289–299.

Dötterl, S. and Vereecken, N. J. (2010) 'The chemical ecology and evolution of bee–flower interactions: a review and perspectives', *Canadian Journal of Zoology*, 88(7), pp. 668–697.

Duraimurugan, P. and Regupathy, A. (2005) 'Stimulo-deterrent diversionary strategy with conjunctive use of trap crops, neem and Bacillus thuringiensis Berliner for the management of insecticide resistant Helicoverpa armigera (Hubner) in cotton'.

Egamberdieva, D. *et al.* (2017) 'Phytohormones and beneficial microbes: essential components for plants to balance stress and fitness', *Frontiers in Microbiology*, 8, p. 2104.

El-Sayed, A. M. *et al.* (2009) 'Potential of "lure and kill" in long-term pest management and eradication of invasive species', *Journal of Economic Entomology*, 102(3), pp. 815–835.

Erb, M. and Reymond, P. (2019) 'Molecular interactions between plants and insect herbivores', *Annual Review of Plant Biology*, 70, pp. 527–557.

Erbilgin, N. *et al.* (2006) 'Exogenous application of methyl jasmonate elicits defenses in Norway spruce (Picea abies) and reduces host colonization by the bark beetle Ips typographus', *Oecologia*, 148(3), pp. 426–436.

Farmer, E. E. and Ryan, C. A. (1990) 'Interplant communication: airborne methyl jasmonate induces synthesis of proteinase inhibitors in plant leaves', *Proceedings of the National Academy of Sciences of the United States of America*, 87(19), pp. 7713–7716.

Fässler, E. *et al.* (2010) 'Effects of indole-3-acetic acid (IAA) on sunflower growth and heavy metal uptake in combination with ethylene diamine disuccinic acid (EDDS)', *Chemosphere*, 80(8), pp. 901–907.

Finkelstein, R. (2013) 'Abscisic acid synthesis and response', *The Arabidopsis Book/American Society of Plant Biologists*, 11, p. e0166.

Gaffney, T. *et al.* (1993) 'Requirement of salicylic acid for the induction of systemic acquired resistance', *Science*, 261(August 1993), pp. 754–757.

Garcia-Brugger, A. *et al.* (2006) 'Early signaling events induced by elicitors of plant defenses', *Molecular Plant-Microbe Interactions*, 19(7), pp. 711–724.

Gols, R. *et al.* (2003) 'Induction of direct and indirect plant responses by jasmonic acid, low spider mite densities, or a combination of jasmonic acid treatment and spider mite infestation', *Journal of Chemical Ecology*, 29(12), pp. 2651–2666.

Gols, R., Posthumus, M. A. and Dicke, M. (1999) 'Jasmonic acid induces the production of gerbera volatiles that attract the biological control agent Phytoseiulus persimilis', *Entomologia Experimentalis et Applicata*, 93(1), pp. 77–86.

Gómez-Cadenas, A. *et al.* (2002) 'Abscisic acid reduces leaf abscission and increases salt tolerance in citrus plants', *Journal of Plant Growth Regulation*, 21, pp. 234–240.

Graniti, A. (1991) 'Phytotoxins and their involvement in plant diseases. Introduction', *Experientia*, 47(8), pp. 751–755.

Grant, M. and Lamb, C. (2006) 'Systemic immunity', *Current Opinion in Plant Biology*, 9(4), pp. 414–420.

Gull, A., Lone, A. A. and Wani, N. U. I. (2019) 'Biotic and abiotic stresses in plants', *Abiotic and biotic stress in plants*, pp. 1–19.

Gupta, R. and Bar, M. (2020) 'Plant immunity, priming, and systemic resistance as mechanisms for Trichoderma spp. biocontrol', *Trichoderma: host pathogen interactions and applications*, pp. 81–110.

Hahn, A. *et al.* (2008) 'The root cap determines ethylene-dependent growth and development in maize roots', *Molecular Plant*, 1(2), pp. 359–367.

Hare, J. D. (2007) 'Variation in herbivore and methyl jasmonate-induced volatiles among genetic lines of Datura wrightii', *Journal of Chemical Ecology*, 33(11), pp. 2028–2043. doi: 10.1007/s10886-007-9375-1.

Hayat, S. *et al.* (2019) *Brassinosteroids: plant growth and development*. Springer.

Hayat, S., Ali, B. and Ahmad, A. (2007) 'Salicylic acid: biosynthesis, metabolism and physiological role in plants', *Salicylic acid: a plant hormone*, pp. 1–14.

He, Z. *et al.* (2000) 'Perception of brassinosteroids by the extracellular domain of the receptor kinase BRI1', *Science*, 288(5475), pp. 2360–2363.

Hedden, P. (2020) 'The current status of research on gibberellin biosynthesis', *Plant and Cell Physiology*, 61(11), pp. 1832–1849.

Hegde, M. *et al.* (2012) 'Aphid antixenosis in cotton is activated by the natural plant defence elicitor cis-jasmone', *Phytochemistry*, 78(2012), pp. 81–88. doi: 10.1016/j.phytochem.2012.03.004.

Heijari, J. *et al.* (2008) 'Long-term effects of exogenous methyl jasmonate application on Scots pine (Pinus sylvestris) needle chemical defence and diprionid sawfly performance', *Entomologia Experimentalis et Applicata*, 128(1), pp. 162–171. doi: 10.1111/j.1570-7458.2008.00708.x.

Hilker, M. and Meiners, T. (2010) 'How do plants "notice" attack by herbivorous arthropods?', *Biological Reviews*, 85(2), pp. 267–280. doi: 10.1111/j.1469-185X.2009.00100.x.

Hogenhout, S. A. and Bos, J. I. B. (2011) 'Effector proteins that modulate plant-insect interactions', *Current Opinion in Plant Biology*, 14(4), pp. 422–428. doi: 10.1016/j.pbi.2011.05.003.

Howe, G. A. and Jander, G. (2008) 'Plant immunity to insect herbivores', *Annual Review of Plant Biology*, 59, pp. 41–66. doi: 10.1146/annurev.arplant.59.032607.092825.

Hudeček, M. *et al.* (2023) 'Plant hormone cytokinin at the crossroads of stress priming and control of photosynthesis', *Frontiers in Plant Science*, 13, p. 1103088.

Hussain, M. A. *et al.* (2020) 'Multifunctional role of brassinosteroid and its analogues in plants', *Plant Growth Regulation*, 92, pp. 141–156.

Ioriatti, C. *et al.* (2011) 'Chemical ecology and management of lobesia botrana (Lepidoptera: Tortricidae)', *Journal of Economic Entomology*, 104(4), pp. 1125–1137. doi: 10.1603/EC10443.

Iqbal, M. and Ashraf, M. (2013) 'Alleviation of salinity-induced perturbations in ionic and hormonal concentrations in spring wheat through seed preconditioning in synthetic auxins', *Acta Physiologiae Plantarum*, 35, pp. 1093–1112.

Iqbal, M., Ashraf, M. and Jamil, A. (2006) 'Seed enhancement with cytokinins: changes in growth and grain yield in salt stressed wheat plants', *Plant Growth Regulation*, 50, pp. 29–39.

Jiang, D. and Yan, S. (2018) 'MeJA is more effective than JA in inducing defense responses in Larix olgensis', *Arthropod-Plant Interactions*, 12(1), pp. 49–56.

Johnson, S. N. and Gregory, P. J. (2006) 'Chemically-mediated host-plant location and selection by root-feeding insects', *Physiological Entomology*, 31(1), pp. 1–13.

Jurado, S. *et al.* (2010) 'The Arabidopsis cell cycle F-box protein SKP2A binds to auxin', *The Plant Cell*, 22(12), pp. 3891–3904.

Kamala Jayanthi, P. D., Raghava, T. and Kempraj, V. (2020) 'Functional diversity of infochemicals in agri-ecological networks', *Innovative pest management approaches for the 21st century: harnessing automated unmanned technologies*, pp. 187–208.

Kansman, J. T. *et al.* (2023) 'Chemical ecology in conservation biocontrol: new perspectives for plant protection', *Trends in Plant Science*, 28, pp. 1166–1177.

Kendrick, M. D. and Chang, C. (2008) 'Ethylene signaling: new levels of complexity and regulation', *Current Opinion in Plant Biology*, 11(5), pp. 479–485.

Khan, A. L. *et al.* (2014) 'Bacterial endophyte Sphingomonas sp. LK11 produces gibberellins and IAA and promotes tomato plant growth', *Journal of Microbiology*, 52, pp. 689–695.

Khan, M. I. R. *et al.* (2015) 'Salicylic acid-induced abiotic stress tolerance and underlying mechanisms in plants', *Frontiers in Plant Science*, 6, p. 462.

Khan, Z. *et al.* (2012) 'Push–pull technology: a conservation agriculture approach for integrated management of insect pests, weeds and soil health in Africa: UK Government's foresight food and farming futures project', *Sustainable intensification*. Routledge, pp. 162–170.

Khan, Z. *et al.* (2016) 'Push-pull: chemical ecology-based integrated pest management technology', *Journal of Chemical Ecology*, 42, pp. 689–697.

Khan, Z. R. *et al.* (2008) 'Chemical ecology and conservation biological control', *Biological Control*, 45(2), pp. 210–224.

Khripach, V. A., Zhabinskii, V. N. and de Groot, A. E. (1998) *Brassinosteroids: a new class of plant hormones*. Academic Press.

Koch, T., Bandemer, K. and Boland, W. (1997) 'Biosynthesis of cis-jasmone: a pathway for the inactivation and the disposal of the plant stress hormone jasmonic acid to the gas phase?', *Helvetica Chimica Acta*, 80(3), pp. 838–850.

Komala, G., Manda, R. R. and Seram, D. (2021) 'Role of semiochemicals in integrated pest management', *International Journal of Entomology Research*, 6(2), pp. 247–253.

Kosakivska, I. V. *et al.* (2022) 'Exogenous phytohormones in the regulation of growth and development of cereals under abiotic stresses', *Molecular Biology Reports*, 49(1), pp. 617–628.

Krishna, P. (2003) 'Brassinosteroid-mediated stress responses', *Journal of Plant Growth Regulation*, 22, pp. 289–297.

Kumawat, P. K., Reena, T. H., Jamwal, S., Sinha, B. K. and Yadav, P. K. (2021) 'Harnessing chemical ecology to address agricultural pest and pollinator: A review'.

Leannec-Rialland, V. *et al.* (2022) 'Use of defensins to develop eco-friendly alternatives to synthetic fungicides to control phytopathogenic fungi and their mycotoxins', *Journal of Fungi*, 8(3), p. 229.

Lefevere, H., Bauters, L. and Gheysen, G. (2020) 'Salicylic acid biosynthesis in plants', *Frontiers in Plant Science*, 11, p. 338.

Li, S.-J. *et al.* (2013) 'Root and shoot jasmonic acid induced plants differently affect the performance of Bemisia tabaci and its parasitoid Encarsia formosa', *Basic and Applied Ecology*, 14(8), pp. 670–679. doi: 10.1016/j.baae.2013.08.011.

Liang, B. *et al.* (2022) 'Salicylic acid is required for broad-spectrum disease resistance in rice', *International Journal of Molecular Sciences*, 23(3), p. 1354.

Loake, G. and Grant, M. (2007) 'Salicylic acid in plant defence-the players and protagonists', *Current Opinion in Plant Biology*, 10(5), pp. 466–472. doi: 10.1016/j.pbi.2007.08.008.

Lomate, P. R. and Hivrale, V. K. (2012) 'Wound and methyl jasmonate induced pigeon pea defensive proteinase inhibitor has potency to inhibit insect digestive proteinases', *Plant Physiology and Biochemistry*, 57, pp. 193–199.

Lou, Y. *et al.* (2005) 'Exogenous application of jasmonic acid induces volatile emissions in rice and enhances parasitism of Nilaparvata lugens eggs by the parasitoid Anagrus nilaparvatae', *Journal of Chemical Ecology*, 31(9), pp. 1985–2002. doi: 10.1007/s10886-005-6072-9.

Malamy, J. *et al.* (1990) 'Salicylic acid: a likely endogenous signal in the resistance response of tobacco to viral infection', *Science*, 250(22), pp. 6–9.

Mamay, M. and Mutlu, Ç. (2019) 'Trend biotechnological management methods against agricultural pests: mating disruption, mass trapping and attract & kill', *1st international Gobeklitepe agriculture congress*, pp. 511–517.

Manghwar, H. *et al.* (2022) 'Brassinosteroids (BRs) role in plant development and coping with different stresses', *International Journal of Molecular Sciences*, 23(3), p. 1012.

Manjili, F. A., Sedghi, M. and Pessarakli, M. (2012) 'Effects of phytohormones on proline content and antioxidant enzymes of various wheat cultivars under salinity stress', *Journal of Plant Nutrition*, 35(7), pp. 1098–1111.

Márquez, G., Alarcón, M. V. and Salguero, J. (2016) 'Differential responses of primary and lateral roots to indole-3-acetic acid, indole-3-butyric acid, and 1-naphthaleneacetic acid in maize seedlings', *Biologia Plantarum*, 60(2), pp. 367–375.

Martin, D. M., Gershenzon, J. and Bohlmann, J. (2003) 'Induction of volatile terpene biosynthesis and diurnal emission by methyl jasmonate in foliage of norway spruce', *Plant Physiology*, 132, pp. 1586–1599. doi: 10.1104/pp.103.021196.some.

Matthes, M. C. *et al.* (2010) 'The transcriptome of cis-jasmone-induced resistance in Arabidopsis thaliana and its role in indirect defence', *Planta*, 232(5), pp. 1163–1180.

Mbaluto, C. M. *et al.* (2020) 'Insect chemical ecology: chemically mediated interactions and novel applications in agriculture', *Arthropod-plant Interactions*, 14, pp. 671–684.

Meena, M. *et al.* (2022) 'Role of elicitors to initiate the induction of systemic resistance in plants to biotic stress', *Plant Stress*, 5, p. 100103.

Méndez-Bravo, A. *et al.* (2019) 'CONSTITUTIVE TRIPLE RESPONSE1 and PIN2 act in a coordinate manner to support the indeterminate root growth and meristem cell proliferating activity in Arabidopsis seedlings', *Plant Science*, 280, pp. 175–186.

Méndez-Bravo, A. *et al.* (2023) 'Beneficial effects of selected rhizospheric and endophytic bacteria, inoculated individually or in combination, on non-native host plant development', *Rhizosphere*, 26, p. 100693.

Miller, J. R. and Cowles, R. S. (1990) 'Stimulo-deterrent diversion: a concept and its possible application to onion maggot control', *Journal of Chemical Ecology*, 16, pp. 3197–3212.

Mohamed, H. I., El-Shazly, H. H. and Badr, A. (2020) 'Role of salicylic acid in biotic and abiotic stress tolerance in plants', *Plant phenolics in sustainable agriculture: volume 1*, pp. 533–554.

Mora-Herrera, M. E. and Lopez-Delgado, H. A. (2007) 'Freezing tolerance and antioxidant activity in potato microplants induced by abscisic acid treatment', *American Journal of Potato Research*, 84, pp. 467–475.

Murphy, S. M. and Loewy, K. J. (2015) 'Trade-offs in host choice of an herbivorous insect based on parasitism and larval performance', *Oecologia*, 179, pp. 741–751.

Naveed, Z. A. *et al.* (2020) 'The PTI to ETI continuum in Phytophthora-plant interactions', *Frontiers in Plant Science*, 11, p. 593905.

Negi, S. *et al.* (2010) 'Genetic dissection of the role of ethylene in regulating auxin-dependent lateral and adventitious root formation in tomato', *The Plant Journal*, 61(1), pp. 3–15.

Nguyen, D. *et al.* (2016) 'How plants handle multiple stresses: hormonal interactions underlying responses to abiotic stress and insect herbivory', *Plant Molecular Biology*, 91, pp. 727–740.

Nolan, T. M. *et al.* (2020) 'Brassinosteroids: multidimensional regulators of plant growth, development, and stress responses', *The Plant Cell*, 32(2), pp. 295–318.

Norbury, G. *et al.* (2014) 'Pest fencing or pest trapping: a bio-economic analysis of cost-effectiveness', *Austral Ecology*, 39(7), pp. 795–807.

Odintsova, T. I. *et al.* (2019) 'Defensin-like peptides in wheat analyzed by whole-transcriptome sequencing: a focus on structural diversity and role in induced resistance', *PeerJ*, 7, p. e6125.

Paque, S. and Weijers, D. (2016) 'Q&A: Auxin: the plant molecule that influences almost anything', *BMC Biology*, 14, pp. 1–5.

Park, S.-W. *et al.* (2007) 'Methyl salicylate is a critical mobile signal for plant systemic acquired resistance', *Science*, 318(5847), pp. 113–116.

Peleg, Z. and Blumwald, E. (2011) 'Hormone balance and abiotic stress tolerance in crop plants', *Current Opinion in Plant Biology*, 14(3), pp. 290–295.

Peñaflor, M. and Bento, J. M. S. (2013) 'Herbivore-induced plant volatiles to enhance biological control in agriculture', *Neotropical Entomology*, 42(4), pp. 331–343.

Pichersky, E. and Gershenzon, J. (2002) 'The formation and function of plant volatiles: perfumes for pollinator attraction and defense', *Current Opinion in Plant Biology*, 5(3), pp. 237–243. doi: 10.1016/S1369-5266(02)00251-0.

Pickett, J. A. (1989) 'Semiochemicals for aphid control', *Journal of Biological Education*, 23(3), pp. 180–186.

Pickett, J. A. *et al.* (1992) 'The chemical ecology of aphids', *Annual Review of Entomology*, 37(1), pp. 67–90.

Pickett, J. A. *et al.* (2007a) 'Developments in aspects of ecological phytochemistry: the role of cis-jasmone in inducible defence systems in plants', *Phytochemistry*, 68(22–24), pp. 2937–2945. doi: 10.1016/j.phytochem.2007.09.025.

Pickett, J. A. *et al.* (2007b) 'Plant volatiles yielding new ways to exploit plant defence', *Chemical ecology*, pp. 161–173. doi: 10.1007/978-1-4020-5369-6_11.

Pickett, J. A. *et al.* (2012) 'Aspects of insect chemical ecology: exploitation of reception and detection as tools for deception of pests and beneficial insects', *Physiological Entomology*, 37(1), pp. 2–9.

Pickett, J. A. and Glinwood, R. T. (July 2007) *Chemical Ecology*, 64(2), pp. 149–156.

Pickett, J. A., Wadhams, L. J. and Woodcock, C. M. (1997) 'Developing sustainable pest control from chemical ecology', *Agriculture, Ecosystems & Environment*, 64(2), pp. 149–156.

Pierre, S. P. *et al.* (2013) 'Belowground induction by *Delia radicum* or phytohormones affect aboveground herbivore communities on field-grown broccoli', *Frontiers in Plant Science*, 4(Aug), p. 305. doi: 10.3389/fpls.2013.00305.

Pieterse, C. M. J. *et al.* (2012) 'Hormonal modulation of plant immunity', *Annual Review of Cell and Developmental Biology*, 28, pp. 489–521. doi: 10.1146/annurev-cellbio-092910-154055.

Pirk, C. W. W. (2021) 'Exploring the kairomone-based foraging behaviour of natural enemies to enhance biological control: a review', *Frontiers in Ecology and Evolution*, 9, p. 143.

Prota, N., Bouwmeester, H. J. and Jongsma, M. A. (2014) 'Comparative antifeedant activities of polygodial and pyrethrins against whiteflies (Bemisia tabaci) and aphids (Myzus persicae)', *Pest Management Science*, 70(4), pp. 682–688.

Raskin, I. (1992) 'Role of salicylic acid in plants', *Annual Review of Plant Biology*, 43(1), pp. 439–463.

Reddy, G. V. P. and Guerrero, A. (2010) 'New pheromones and insect control strategies', *Vitamins & Hormones*, 83, pp. 493–519.

Renwick, J. A. A. (2002) 'The chemical world of crucivores: lures, treats and traps', *Proceedings of the 11th international symposium on insect-plant relationships*. Springer, pp. 35–42.

Rivero, R. M. *et al.* (2010) 'Enhanced cytokinin synthesis in tobacco plants expressing PSARK:: IPT prevents the degradation of photosynthetic protein complexes during drought', *Plant and Cell Physiology*, 51(11), pp. 1929–1941.

Rivero, R. M., Shulaev, V. and Blumwald, E. (2009) 'Cytokinin-dependent photo-respiration and the protection of photosynthesis during water deficit', *Plant Physiology*, 150(3), pp. 1530–1540.

Robles, L. M. *et al.* (2012) 'A loss-of-function mutation in the nucleoporin AtNUP160 indicates that normal auxin signalling is required for a proper ethylene response in Arabidopsis', *Journal of Experimental Botany*, 63(5), pp. 2231–2241.

Rodriguez-Saona, C. *et al.* (2001) 'Exogenous methyl jasmonate induces volatile emissions in cotton plants', *Journal of Chemical Ecology*, 27(4), pp. 679–695.

Rodriguez-Saona, C., Blaauw, B. R. and Isaacs, R. (2012) 'Manipulation of natural enemies in agroecosystems: habitat and semiochemicals for sustainable insect pest control', *Integrated pest management and pest control–current and future tactics*, pp. 89–126.

Rodriguez-Saona, C. R., Polashock, J. and Malo, E. A. (2013) 'Jasmonate-mediated induced volatiles in the american cranberry, Vaccinium macrocarpon: from gene expression to organismal interactions', *Frontiers in Plant Science*, 4(April), p. 115. doi: 10.3389/fpls.2013.00115.

Ryan, C. A. (2000) 'The systemin signaling pathway: differential activation of plant defensive genes', *Biochimica et Biophysica Acta (BBA)-Protein Structure and Molecular Enzymology*, 1477(1–2), pp. 112–121.

Sachs, T. (2005) 'Auxin's role as an example of the mechanisms of shoot/root relations', *Plant and Soil*, 268, pp. 13–19.

Salam, A. *et al.* (2022) 'Nano-priming against abiotic stress: a way forward towards sustainable agriculture', *Sustainability*, 14(22), p. 14880.

Sarkar, S. C. *et al.* (2018) 'Application of trap cropping as companion plants for the management of agricultural pests: a review', *Insects*, 9(4), p. 128.

Schiebe, C. *et al.* (2012) 'Inducibility of chemical defenses in Norway spruce bark is correlated with unsuccessful mass attacks by the spruce bark beetle', *Oecologia*, 170(1), pp. 183–198. doi: 10.1007/s00442-012-2298-8.

Schmülling, T. (2002) 'New insights into the functions of cytokinins in plant development', *Journal of Plant Growth Regulation*, 21(1), pp. 40–49.

Semiz, G. *et al.* (2012) 'Manipulation of VOC emissions with methyl jasmonate and carrageenan in the evergreen conifer Pinus sylvestris and evergreen broadleaf Quercus ilex', *Plant Biology*, 14(SUPPL. 1), pp. 57–65. doi: 10.1111/j.1438-8677.2011.00485.x.

Senaratna, T. *et al.* (2000) 'Acetyl salicylic acid (Aspirin) and salicylic acid induce multiple stress tolerance in bean and tomato plants', *Plant Growth Regulation*, 30(2), pp. 157–161.

Seo, H. S. *et al.* (2001) 'Jasmonic acid carboxyl methyltransferase: a key enzyme for jasmonate-regulated plant responses', *Proceedings of the National Academy of Sciences of the United States of America*, 98(8), pp. 4788–4793. doi: 10.1073/pnas.081557298.

Singewar, K., Fladung, M. and Robischon, M. (2021) 'Methyl salicylate as a signaling compound that contributes to forest ecosystem stability', *Trees*, 35, pp. 1755–1769.

Sobhy, I. S. *et al.* (2017) 'cis-Jasmone elicits aphid-induced stress signalling in potatoes', *Journal of Chemical Ecology*, 43(1), pp. 39–42. doi: 10.1007/s10886-016-0805-9.

Sobhy, I. S., Lou, Y. and Bruce, T. J. A. (2022) *Inducing plant resistance against insects using exogenous bioactive chemicals: key advances and future perspectives*. Frontiers Media SA.

Sondheimer, E. (2012) *Chemical ecology*. Elsevier.

Sosa, M. E. and Tonn, C. E. (2008) 'Plant secondary metabolites from Argentinean semiarid lands: bioactivity against insects', *Phytochemistry Reviews*, 7, pp. 3–24.

Sosnowski, J., Truba, M. and Vasileva, V. (2023) 'The impact of auxin and cytokinin on the growth and development of selected crops', *Agriculture*, 13(3), p. 724.

Spears, L. R. *et al.* (2016) 'Pheromone lure and trap color affects bycatch in agricultural landscapes of Utah', *Environmental Entomology*, 45(4), pp. 1009–1016.

Stella de Freitas, T. F., Stout, M. J. and Sant'Ana, J. (2019) 'Effects of exogenous methyl jasmonate and salicylic acid on rice resistance to Oebalus pugnax', *Pest Management Science*, 75(3), pp. 744–752.

Stepanova, A. N. *et al.* (2007) 'Multilevel interactions between ethylene and auxin in Arabidopsis roots', *The Plant Cell*, 19(7), pp. 2169–2185.

Stotz, H. U., Thomson, J. and Wang, Y. (2009) 'Plant defensins: defense, development and application', *Plant Signaling & Behavior*, 4(11), pp. 1010–1012.

Street, I. H. *et al.* (2015) 'Ethylene inhibits cell proliferation of the Arabidopsis root meristem', *Plant Physiology*, 169(1), pp. 338–350.

Suckling, D. M. *et al.* (2000) 'Pheromones and other semiochemicals', *Biological and biotechnological control of insect pests*, pp. 63–99.

Sunamura, E. *et al.* (2011) 'Combined use of a synthetic trail pheromone and insecticidal bait provides effective control of an invasive ant', *Pest Management Science*, 67(10), pp. 1230–1236.

Swami, V. P. *et al.* (2022) 'Push-Pull strategies in integrated pest management', 9(8).

Swarup, R. *et al.* (2007) 'Ethylene upregulates auxin biosynthesis in Arabidopsis seedlings to enhance inhibition of root cell elongation', *The Plant Cell*, 19(7), pp. 2186–2196.

Tan, K. H. *et al.* (2014) 'Pheromones, male lures, and trapping of tephritid fruit flies', *Trapping and the detection, control, and regulation of tephritid fruit flies: lures, area-wide programs, and trade implications*, pp. 15–74.

Tanaka, K. *et al.* (2009) 'Highly selective tuning of a silkworm olfactory receptor to a key mulberry leaf volatile', *Current Biology*, 19(11), pp. 881–890. doi: 10.1016/j.cub.2009.04.035.

Tanimoto, E. (2005) 'Regulation of root growth by plant hormones—roles for auxin and gibberellin', *Critical Reviews in Plant Sciences*, 24(4), pp. 249–265.

Tantasawat, P. A., Sorntip, A. and Pornbungkerd, P. (2015) 'Effects of exogenous application of plant growth regulators on growth, yield, and in vitro gynogenesis in cucumber', *HortScience*, 50(3), pp. 374–382.

Thaler, J. S. (1999) 'Induced resistance in agricultural crops: effects of jasmonic acid on herbivory and yield in tomato plants', *Environmental Entomology*, 28(1), pp. 30–37.

Thaler, J. S. *et al.* (2001) 'Jasmonate-mediated induced plant resistance affects a community of herbivores', *Ecological Entomology*, 26(3), pp. 312–324.

Thaler, J. S. *et al.* (2002) 'Jasmonate-deficient plants have reduced direct and indirect defences against herbivores', *Ecology Letters*, 5, pp. 764–774.

Ton, J. *et al.* (2007) 'Priming by airborne signals boosts direct and indirect resistance in maize', *The Plant Journal*, 49(1), pp. 16–26.

Tonğa, A. *et al.* (2020) 'cis-Jasmone treatments affect multiple sucking insect pests and associated predators in cotton', *Entomologia Generalis*, 40, pp. 49–61.

Tripathi, D., Raikhy, G. and Kumar, D. (2019) 'Chemical elicitors of systemic acquired resistance—Salicylic acid and its functional analogs', *Current Plant Biology*, 17, pp. 48–59.

Tuerkkan, M. and Dolar, F. S. (2008) 'Role of phytotoxins in plant diseases', *Journal of Agricultural Sciences-Tarim Bilimleri Dergis*, 14(1), pp. 87–94.

Uefune, M., Ozawa, R. and Takabayashi, J. (2014) 'Prohydrojasmon treatment of lima bean plants reduces the performance of two-spotted spider mites and induces volatiles', *Journal of Plant Interactions*, 9(1), pp. 69–73.

Usman, M. G. *et al.* (2018) 'Plant disease control: understanding the roles of toxins and phytoalexins in host-pathogen interaction', *Pertanika Journal of Scholarly Research Reviews*, 4(1).

Vaishnav, D. and Chowdhury, P. (2023) 'Types and function of phytohormone and their role in stress'.

Vardhini, B. V. *et al.* (2010) 'Role of brassinosteroids in alleviating various abiotic and biotic stresses-a review', *Plant Stress*, 4(1), pp. 55–61.

Verma, V., Ravindran, P. and Kumar, P. P. (2016) 'Plant hormone-mediated regulation of stress responses', *BMC Plant Biology*, 16, pp. 1–10.

Vijayan, P. *et al.* (1998) 'A role for jasmonate in pathogen defense of Arabidopsis', *Proceedings of the National Academy of Sciences*, 95(12), pp. 7209–7214.

Vincent, D., Rafiqi, M. and Job, D. (2020) 'The multiple facets of plant–fungal interactions revealed through plant and fungal secretomics', *Frontiers in Plant Science*, 10, p. 1626.

Walters, D. R., Newton, A. C. and Lyon, G. D. (2014) *Induced resistance for plant defense: a sustainable approach to crop protection*. John Wiley & Sons.

Wang, Y. *et al.* (2007) 'Arabidopsis EIN2 modulates stress response through abscisic acid response pathway', *Plant Molecular Biology*, 64, pp. 633–644.

Wang, Z. *et al.* (2012) 'Defense to Sclerotinia sclerotiorum in oilseed rape is associated with the sequential activations of salicylic acid signaling and jasmonic acid signaling', *Plant Science*, 184, pp. 75–82.

War, A. R. *et al.* (2012) 'Mechanisms of plant defense against insect herbivores', *Plant Signaling and Behavior*, 7(10). doi: 10.4161/psb.21663.

War, A. R. *et al.* (2020) 'Plant defense and insect adaptation with reference to secondary metabolites', *Co-evolution of secondary metabolites*, pp. 795–822.

Wasternack, C. (2007) 'Jasmonates: an update on biosynthesis, signal transduction and action in plant stress response, growth and development', *Annals of Botany*, 100, pp. 681–697. doi: 10.1093/aob/mcm079.

Wasternack, C. (2014) 'Action of jasmonates in plant stress responses and development – Applied aspects', *Biotechnology Advances*, 32(1), pp. 31–39. doi: 10.1016/j.biotechadv.2013.09.009.

Webster, B. *et al.* (2008) 'Identification of volatile compounds used in host location by the black bean aphid, Aphis fabae', *Journal of Chemical Ecology*, 34(9), pp. 1153–1161. doi: 10.1007/s10886-008-9510-7.

Webster, B. and Cardé, R. T. (2017) 'Use of habitat odour by host-seeking insects', *Biological Reviews*, 92(2), pp. 1241–1249.

Wei, X. *et al.* (2021) 'Application of methyl jasmonate and salicylic acid lead to contrasting effects on the plant's metabolome and herbivory', *Plant Science*, 303, p. 110784.

Weinzierl, R. *et al.* (2005) 'Insect attractants and traps', *ENY277*. Avalaible: http://ufdcimages.uflib.ufl.edu/IR/00/00/27/94/00001/IN08000.pdf [Accessed 23 Maret 2015].

White, R. F. (1979) 'Acetylsalicylic acid (aspirin) induces resistance to tobacco mosaic virus in tobacco', *Virology*, 99(2), pp. 410–412. doi: 10.1016/0042-6822(79)90019-9.

Woodward, A. W. and Bartel, B. (2005) 'Auxin: regulation, action, and interaction', *Annals of Botany*, 95(5), pp. 707–735.

Wu, H. *et al.* (2011) 'RAC/ROP GTPases and auxin signaling', *The Plant Cell*, 23(4), pp. 1208–1218.

Wyatt, T. D. (2014) *Pheromones and animal behavior: chemical signals and signatures.* Cambridge University Press.

Xia, X.-J. *et al.* (2009) 'Reactive oxygen species are involved in brassinosteroid-induced stress tolerance in cucumber', *Plant Physiology*, 150(2), pp. 801–814.

Xia, X. *et al.* (2011) 'Induction of systemic stress tolerance by brassinosteroid in Cucumis sativus', *New Phytologist*, 191(3), pp. 706–720.

Xiong, J. *et al.* (2020) 'Brassinosteroids are involved in ethylene-induced Pst DC3000 resistance in Nicotiana benthamiana', *Plant Biology*, 22(2), pp. 309–316.

Yamaguchi, S. (2008) 'Gibberellin metabolism and its regulation', *Annual Review of Plant Biology*, 59, pp. 225–251.

Yang, J. *et al.* (2019) 'The crosstalks between jasmonic acid and other plant hormone signaling highlight the involvement of jasmonic acid as a core component in plant response to biotic and abiotic stresses', *Frontiers in Plant Science*, 10, p. 1349.

Yang, Y.-X. *et al.* (2015) 'Crosstalk among jasmonate, salicylate and ethylene signaling pathways in plant disease and immune responses', *Current Protein and Peptide Science*, 16(5), pp. 450–461.

Yao, H. and Tian, S. (2005) 'Effects of pre-and post-harvest application of salicylic acid or methyl jasmonate on inducing disease resistance of sweet cherry fruit in storage', *Postharvest Biology and Technology*, 35(3), pp. 253–262.

Yew, J. Y. and Chung, H. (2015) 'Insect pheromones: an overview of function, form, and discovery', *Progress in Lipid Research*, 59, pp. 88–105.

Yu, M.-H., Zhao, Z.-Z. and He, J.-X. (2018) 'Brassinosteroid signaling in plant–microbe interactions', *International Journal of Molecular Sciences*, 19(12), p. 4091.

Zhang, P. *et al.* (2010) 'Senescence-inducible expression of isopentenyl transferase extends leaf life, increases drought stress resistance and alters cytokinin metabolism in cassava', *Journal of Integrative Plant Biology*, 52(7), pp. 653–669.

Zhang, P. J. *et al.* (2013) 'Feeding by whiteflies suppresses downstream jasmonic acid signaling by eliciting salicylic acid signaling', *Journal of Chemical Ecology*, 39(5), pp. 612–619. doi: 10.1007/s10886-013-0283-2.

Zhang, Y. *et al.* (2022) 'Molecular mechanisms of diverse auxin responses during plant growth and development', *International Journal of Molecular Sciences*, 23(20), p. 12495.

Zhao, B. *et al.* (2021) 'Roles of phytohormones and their signaling pathways in leaf development and stress responses', *Journal of Agricultural and Food Chemistry*, 69(12), pp. 3566–3584.

Zhao, T. *et al.* (2011) 'Host resistance elicited by methyl jasmonate reduces emission of aggregation pheromones by the spruce bark beetle, Ips typographus', *Oecologia*, 167(3), pp. 691–699. doi: 10.1007/s00442-011-2017-x.

Zhu, F. *et al.* (2014) 'Salicylic acid and jasmonic acid are essential for systemic resistance against tobacco mosaic virus in Nicotiana benthamiana', *Molecular Plant-Microbe Interactions*, 27(6), pp. 567–577.

Zipfel, C. (2009) 'Early molecular events in PAMP-triggered immunity', *Current Opinion in Plant Biology*, 12(4), pp. 414–420. doi: 10.1016/j.pbi.2009.06.003.

Zipfel, C. and Robatzek, S. (2010) 'Pathogen-associated molecular pattern-triggered immunity: veni, vidi...?', *Plant Physiology*, 154(2), pp. 551–554.

CHAPTER 8

Techniques Used in Chemical Ecology

8.1 INTRODUCTION

The field of chemical ecology stands at the intersection of multiple scientific domains, weaving together chemistry, biology and ecology to decipher the intricate language of chemical signals that shape interactions within ecosystems (Bergström, 2007; Hartmann, 2008; Mithöfer, Boland and Maffei, 2009). Central to this discipline is the exploration of the diverse mechanisms by which organisms communicate, negotiate and respond to their environments through chemical cues. As chemical ecologists delve into the rich tapestry of these interactions, they rely on a spectrum of techniques that enable them to decode the complex dialogues that unfold among species (Dicke and van Loon, 2014; Meiners, 2015). The world of chemical ecology teems with wonder as researchers employ these techniques to investigate how organisms perceive and respond to the intricate symphony of chemical cues that envelop them. By studying the physiological, behavioural and electrophysiological responses triggered by these signals, we gain insights into the strategies that organisms employ to navigate their environments, find mates, locate resources and defend against threats (Millar and Haynes, 1998; Dyer et al., 2018). As we journey through the chapters of this book, the techniques presented here will serve as our compass, guiding us through the labyrinth of chemical interactions that shape the natural world.

DOI: 10.1201/9781003479857-8

The study of chemical ecology focuses on understanding the chemical interactions between organisms in their environments. One of the most common and fascinating examples of chemical ecology is the interaction between insects and plants (Mbaluto et al., 2020). Insects rely on plants for food and shelter, but they also use chemical cues to locate their host plants, identify potential mates and avoid predators. Understanding the chemical ecology of insect–plant interactions is, therefore, essential for pest management, conservation and ecological research (Mithöfer, Boland and Maffei, 2009; Mbaluto et al., 2020). This area of research seeks to understand the complex chemical communication between insects and plants. The chemical language that insects and plants use to communicate plays a crucial role in shaping the ecology and evolution of the organisms involved. In the past, it was challenging to study insect–plant communication because of the small amounts of chemicals involved, and the difficulties in identifying, isolating and analysing them. However, the development of advanced techniques has made it possible to investigate this field more extensively. In this chapter, we will discuss some of the most commonly used techniques in chemical ecology research, including performance bioassays, behavioural bioassays, entrainment collection, and volatile analysis using gas chromatography-mass spectrometry (GC-MS) and electrophysiology. These techniques allow researchers to study the effects of plant chemicals on insect behaviour, growth, fecundity and mortality and to identify the specific compounds responsible for these effects.

8.2 PERFORMANCE BIOASSAY

In order to assess the performance of insects there are various parameters including growth, fecundity and mortality. These bioassays are performance-based techniques used in the study of insect–plant interactions. These methods are useful in evaluating the impact of treatment on insect performance and behaviour. For example, these bioassays can determine whether the chemicals from a particular plant affect the growth of insects. It can also determine whether the chemicals affect the fecundity and/or induce mortality of the insect populations.

8.2.1 Clip-Cage Bioassay

The use of clip-cages is one of the most common methods to evaluate the performance of insects, as it helps in restricting insect movement to the area of interest for researchers. The goal is to observe the insects' behaviour, such as feeding or mating, in response to stimuli. These clip-cages are specially designed and have proper ventilation to avoid any external stress on insects, ensuring that the results obtained from the experiments are the effect of the treatment (Ali et al., 2021; Ali, Sobhy and Bruce, 2022).

8.2.2 BugDorm Bioassay

BugDorms are larger versions of clip-cages that facilitate research by enclosing entire plants, thus allowing insects to feed on them. These BugDorms offer versatility, accommodating both no-choice and choice bioassays, depending on the research design. For instance, various studies have adopted BugDorms for settlement and foraging bioassays, as demonstrated by Ali et al. (2021). In these bioassays, four plants (two treated and two control) were enclosed within the BugDorm, providing insects with more freedom compared to clip-cages. This method has been widely utilised in insect–plant interaction studies (Ali et al., 2024).

8.3 BEHAVIOURAL BIOASSAY

8.3.1 Wind Tunnel Bioassay

The wind tunnel bioassay is a laboratory technique that is used to study the behaviour of insects and other small organisms in simulated wind environments. It is a tool that helps researchers understand how insects navigate and respond to wind stimuli, which is important in developing pest management strategies (Figure 8.1). The wind tunnel bioassay typically consists of a long, clear glass or plastic tube that is enclosed on all sides, except for the two open ends. The tube has a fan or blower at one end that creates a laminar airflow through the tube. At the opposite end, there is a small opening where insects can be introduced into the tunnel. The airflow rate can be adjusted to simulate different wind speeds and patterns. To conduct a wind tunnel bioassay, insects are collected and introduced into the tunnel through the small opening. The insects are then observed, and their behaviour is recorded as they move through the tunnel. Researchers can measure various parameters such as flight speed, trajectory and response to different wind stimuli. Different types of wind tunnel bioassays can be used to study various aspects of insect behaviour. For example, a straight-line tunnel bioassay can be used to measure the flight speed and trajectory of insects under different wind conditions. A Y-shaped tunnel bioassay can be used to study insects' response to different odour or visual cues in the presence of wind stimuli. The data collected from wind tunnel bioassays can be used to study the mechanisms of insect

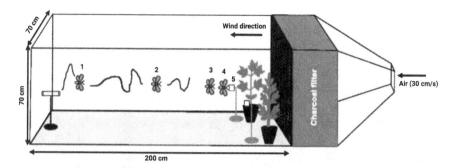

FIGURE 8.1 A diagram depicting the wind tunnel setup. For dual-choice bioassays, a host plant (cotton) and a non-host plant were positioned upwind in the tunnel, maintaining a 20 cm separation between them. To attract male insects, a filter paper loaded with a single female equivalent pheromone blend was placed in front of each plant. Male insects were subsequently released downwind from a glass tube. 1) Take off, 2) half-way), 3) close approach), 4) landing and 5) filter paper (Adapted from Binyameen et al., 2013).

navigation and behaviour. For example, studies have shown that insects use their antennae and other sensory structures to detect wind direction and speed, which helps them navigate over long distances. In conclusion, the wind tunnel bioassay is a powerful laboratory technique that allows researchers to study the behaviour of insects under different wind conditions. The insights gained from this technique can be used to develop more effective pest management strategies.

8.3.2 Olfactometer Bioassay

The sense of smell, or olfaction, plays an essential role in the life of insects, from mating and foraging to avoiding predators and identifying hosts. Olfactometers are scientific instruments used to study insect behaviour in response to various odours, with the aim of understanding the underlying neural mechanisms involved in insect olfaction (Roberts et al., 2023). Olfactometers come in different types, but all share a common function of delivering controlled streams of airborne chemicals to an insect's sensory receptors. The most commonly used olfactometer type is the wind tunnel, which generates a well-defined, measurable airflow that guides the insect's behaviour toward or away from the odour source (Roberts et al., 2023) (Figure 8.2). Other designs include the Y-tube olfactometer, which offers a choice between two odours, and the four-chamber olfactometer, which allows for simultaneous comparisons of multiple odours. The versatility

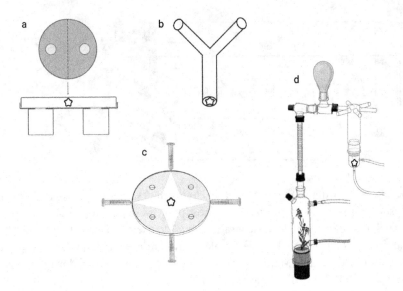

FIGURE 8.2   Illustrations of various olfactometer designs: a) still air olfactometer, b) Y-olfactometer, c) four-arm olfactometer and d) upright six-arm olfactometer. The star indicates the entry point for the study insect.

of olfactometers enables researchers to investigate a wide range of insect behaviours, including host and mate selection, pheromone-mediated communication and responses to plant volatiles, among others. As a result, olfactometry has become an indispensable tool in the fields of entomology, neurobiology and insect pest management.

8.3.2.1 Y-Olfactometer

A Y-olfactometer is a specialised piece of equipment that is frequently used to study insect behaviour. It is comprised of a Y-shaped tube that is divided into two chambers, with each chamber leading to a different stimulus source. Typically, the stimulus sources will be vials of different odours or chemical compounds, which are placed at the ends of each chamber. Insects are then placed in the centre of the Y-tube, and their behaviour is observed as they move toward one of the stimuli. This apparatus allows researchers to measure an insect's attraction or aversion to particular stimuli and can be used to study a wide range of behaviours, such as host plant selection, mating behaviour and feeding preferences. The Y-olfactometer has proven to be a useful tool in understanding the sensory capabilities of insects and can provide vital insights into pest management strategies and ecology.

8.3.2.2 Four-Arm Olfactometer

The four-arm olfactometer apparatus is a sophisticated experimental tool used to study insect behaviour. It consists of a central platform and four arms, each of which can be separately perfused with different stimuli or odours. The insects are placed on the central platform and given a choice to move toward their preferred odour or stimulus in the arms. The movement of the insects toward their favourite stimuli is then recorded, and the data obtained are analysed to understand the behavioural responses of the insects. The apparatus is used for various purposes such as investigating the olfactory preferences of insects, examining the effects of various chemicals on their behaviour and understanding the underlying mechanisms of insect behaviour and sensory perception. The four-arm olfactometer apparatus is a valuable tool for insect behavioural research and has contributed significantly to our understanding of the behavioural ecology of insects. In a four-arm olfactometer, the insect is enclosed in an area divided into four regions, as the name indicates. Each region has an opening for the odour source. In a four-arm olfactometer, total time spent by an insect in each arm is recorded to determine the insects' preference for different odour sources. A four-arm olfactometer does not distinguish between attraction and arrestant or between repulsion and locomotion stimulation because it is designed to record the time spent by the insects; it does not tell how the insect has moved in response to the odour source. In contrast, the four-arm olfactometer has several advantages, including allowing insects to make multiple choices and avoiding initial behaviour responses that may be due to the stress of being placed into an olfactometer. While in a locomotion compensator olfactometer, insects are placed on a sphere, and the odour source is blown over the insect; when the insect moves in response to the odour source, it causes the rotation of the sphere. The sphere's rotation is then recorded and describes the direction of movement, speed of movement and rate of turning of the insects, which allows the recording of different behaviour.

Y-olfactometers and four-arm olfactometers are two types of devices used in sensory analysis to evaluate the sense of smell. The main differences between the two are design, application, precision, uses, and cost.

- *Design*: The Y-olfactometer is designed in a Y-shape, which splits into two arms that hold aroma sources, whereas the four-arm olfactometer is constructed in a square shape, which consists of four arms equipped with aroma sources.

- *Application*: The Y-olfactometer is used for testing individual odours, while the four-arm olfactometer is used for testing multiple odours and their interactions.

- *Precision*: The Y-olfactometer provides a more precise odour measurement, as it allows testers to choose a specific odorant from a set of different strengths. In contrast, the four-arm olfactometer is designed for testing the overall quality of odorants and cannot measure individual odorants as precisely.

- *Uses*: The Y-olfactometer is commonly used in medical and industrial settings, while the four-arm olfactometer is used in the food and beverage industry.

- *Cost*: The Y-olfactometer is more expensive than the four-arm olfactometer because of its complex design and higher precision.

8.3.2.3 Still Air Olfactometer

A still air olfactometer is a device used in chemical ecology and behavioural studies to test the responses of insects or other organisms to specific odours (Prokopy, Cooley and Phelan, 1995). It creates a controlled environment in which the movement of air is minimised or eliminated, allowing the target organisms to detect and respond to odorous cues without the interference of external air currents (Van Tol, Visser and Sabelis, 2002). The setup typically consists of a chamber with separate compartments, often connected by tubing or channels (Siljander et al., 2008). One compartment contains the source of the odour, which can be a volatile chemical compound, plant material or other scented substances. The other compartment holds the test subject, usually an insect (Van Tol, Visser and Sabelis, 2002; Weeks et al., 2011). The key feature of a still air olfactometer is its design to prevent the odour molecules from dissipating rapidly, ensuring that the test organism can perceive and respond to the odour gradient in a stable and controlled manner (Weeks et al., 2011).

8.3.2.4 Upright Six-Arm Olfactometer

An upright six-arm olfactometer is an apparatus commonly used to study insect behaviour and chemical communication. It consists of a central chamber with multiple arms extending outward, each leading to a separate testing arena. The arms are typically transparent and can be fitted with various odorous stimuli or test subjects. This olfactometer is designed to assess the preferences or responses of insects to different odour sources or environmental conditions. In experiments, the insects are placed in the central chamber, and air currents are introduced to carry the odours from the arms into the central chamber. By observing the insects' movement and choices, researchers can determine their attraction, avoidance or neutral response to the different odour cues. The upright six-arm olfactometer provides a controlled and replicable environment for studying chemically mediated behaviours, such as host plant selection, mating preferences or responses to synthetic compounds. Its design allows for simultaneous testing of multiple individuals or species, making it a valuable tool in understanding chemical ecology and insect behaviour.

8.4 ENTRAINMENT COLLECTION

Entrainment collection is a technique used to capture insects for chemical analysis. It involves the use of a light source to attract the insects into a trap that is equipped with a chemical absorbent (Tholl et al., 2021). The entrainment technique is used to obtain a variety of insect specimens for analysis. Headspace sampling allows samples to be collected with the chemicals released by the plant in ratios similar to those found in nature.

During volatile collection, the plant is kept inside a bag (35 x 43 cm; Bacofoil, UK) to collect the plant volatiles (Figures 8.3–8.5). The bag is partially sealed, so that only the volatiles produced by the plants are collected. Prior to entrainments, bags are baked in an oven (Heraeus, Thermo Electron corporation, Mark Biosciences, UK) at 120°C overnight. The Porapak Q filters (0.05 g, 60/80 mesh; Supelco, Bellefonte, PA, USA) are rinsed with diethyl ether and conditioned before use. Plants with five true leaves are enclosed in bags individually. Each bag is open at the bottom and closed at the top. An outlet hole is made in the upper part of the bag to connect the Porapak Q filter, and the bag is closed by attaching a rubber band around the pot. Charcoal-filtered air is pumped in at 600 ml min-1, and sampled air is pulled out at 400 ml min-1 through a Porapak Q filter in which the plant volatiles are trapped. To avoid the entry of unfiltered air, positive pressure is maintained by using differing flows rates. Connections are made with 1.6 mm (i.d.) polytetrafluoroethylene (PTFE) tubing (Alltech Associates Inc., Lancashire, UK) with Swagelok brass ferrules and fittings (North London Valve Co., London, UK) and sealed with PTFE tape (Gibbs & Dandy Ltd., Luton, UK). Volatile collection is done for a period of 48 hours, after which, the Porapak filters are eluted with 0.5 ml of diethyl ether into sample vials (Supelco, 2 ml, PTFE/silicone)

168 ■ Chemical Ecology

Practical approaches to plant volatile analysis

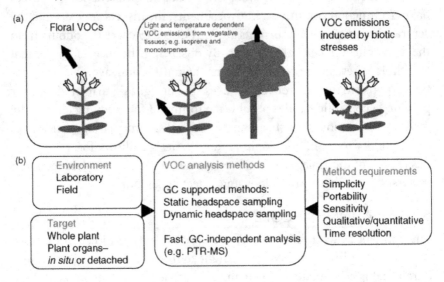

FIGURE 8.3 Strategies for plant volatile analysis: (a) typical sources of plant volitile organic compound (VOC) emissions, (b) considerations for planning VOC analysis experiments (Adapted from Tholl et al., 2021).

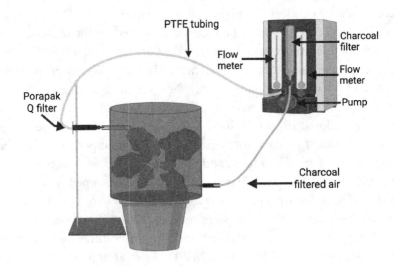

FIGURE 8.4 Illustration showing a diagrammatic representation of plant volatile collection setup.

Practical approaches to plant volatile analysis

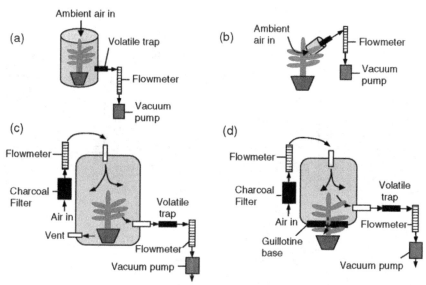

FIGURE 8.5 Examples of dynamic headspace collection systems. (a) In a simple "pull" headspace collection chamber, the plant is enclosed in an open-top container. An air stream, regulated by a flow meter, is pulled over the plant and through a VOC collecting adsorbent trap. (b) A "pull" headspace collection device with an open-top chamber for collecting VOCs from a single leaf. (c) In a "push–pull" headspace collection system, pressurised air enters the top of the collection chamber regulated by a flow meter. Incoming air is purified by passing through a charcoal filter placed behind or in front of the flow meter. Alternatively, high-purity synthetic air may be used. After passing over the plant sample, the air is pulled through an adsorbent volatile trap at the lower side of the chamber at a defined rate controlled by a second flow meter. Excess air can escape through the vent on the lower side of the chamber. (d) Example of a modified "push–pull" headspace collection chamber for collecting VOCs from parts of a plant. Teflon-coated guillotine-like blades close the base of the chamber around the stem of the plant allowing VOCs from the upper part of the plant to be trapped (Adapted from Tholl et al., 2006).

and stored at -20°C in a freezer (Lec Medical, UK) for use in olfactometer bioassays and chemical analysis.

8.5 VOLATILE ANALYSIS

GC-MS is a widely used method for analysing volatile chemicals from plants involved in insect–plant interactions (Figure 8.6). GC-MS is used to identify and quantify the chemicals that are released by plants. In many

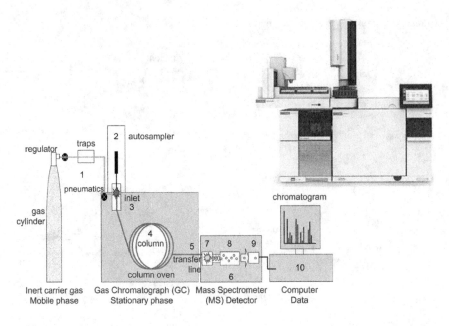

FIGURE 8.6   Illustration showing the sequential steps involved in GC-MS analysis from injection of the prepared sample into the GC instrument to the generation of a chromatogram.

cases, GC-MS is used to determine the dose of chemicals needed to attract or deter insects. GC-MS utilises the combination of two methods, namely gas chromatography and mass spectrometry for the identification and quantification of volatile compounds present in samples. This analytical technique has become one of the most widely used methods in the field of organic chemistry, biochemistry and environmental sciences for the detection of chemical compounds that may be present in various matrices, including plant samples. Gas chromatography involves the separation of volatile compounds in a mixture based on their affinity toward the stationary phase in a chromatography column. The sample is injected into the column, and a gas such as helium is used as the mobile phase to carry the sample through the column. As the sample passes through the column, each volatile compound in the sample interacts differently with the stationary phase, resulting in different retention times or elution times. This process separates the volatile compounds in the sample and creates a chromatogram, which is a graphical representation of the separated compounds.

Mass spectrometry, on the other hand, involves the measurement of the mass-to-charge ratio (m/z) of ions generated from the fragmentation of volatile compounds in a sample. The separated compounds from the gas chromatography column pass through a mass spectrometer, which ionises the compounds and fragments them into smaller molecules. These fragments are then separated by their m/z, creating a mass spectrum. The fragmentation pattern produced is unique for each compound, which allows for their identification. The combination of these two techniques, gas chromatography and mass spectrometry, provides an enhanced level of sensitivity and selectivity, making GC-MS the preferred technique for the identification of volatile compounds in complex mixtures such as plant samples. The use of GC-MS in the analysis of plant samples has gained significant attention, particularly in the fields of pharmacognosy, phytochemistry and natural products research. These applications have led to the identification of various bioactive compounds in plant samples, which can be used for the discovery of new drugs and development of botanical products for therapeutic purposes. In summary, GC-MS is a powerful analytical technique that combines gas chromatography and mass spectrometry to analyse and identify volatile compounds in complex mixtures. The technique plays a crucial role in the identification and quantification of volatile compounds in plant samples, which has a significant impact on the fields of pharmacognosy, phytochemistry and natural products research.

8.6 ELECTROPHYSIOLOGY

Gas chromatography-electroantennography (GC-EAG) is a powerful analytical tool used in chemical ecology to study insect behaviour. It involves the extraction and isolation of volatile compounds from biological samples such as plants, insects and their habitats, followed by the use of gas chromatography to separate and identify the compounds. The extracted compounds are then presented to the antennae of the insect, and their responses are recorded using an electroantennogram. The separated compounds are then presented to the antennae of the insect, and their responses are recorded using an electroantennogram. The importance of GC-EAG in studying insect behaviour in chemical ecology lies in its ability to detect and identify the volatile compounds that play a vital role in the chemical communication between insects and their environments. These compounds are often involved in insect mating, foraging, host location and other critical behaviours. Studying the responses of insect antennae to these compounds can help us to understand their behaviour and ecology and, ultimately, develop strategies for controlling insect pests and diseases.

8.7 CONCLUSION

The techniques discussed in this chapter are vital in defining and understanding the chemical ecology in insect–plant interactions. They have revolutionised our understanding of insect behaviour, including the selection of host plants, and the physiological effects of plant chemicals on insects. Furthermore, the techniques outlined in this chapter have contributed significantly to understanding the mechanisms of insect–plant communication. They have also given us valuable insights that can be used in developing pest management strategies and improving crop yields.

REFERENCES

Ali, J. *et al.* (2021) 'Effects of cis-Jasmone treatment of brassicas on interactions with Myzus persicae Aphids and their parasitoid diaeretiella rapae', *Frontiers in Plant Science*, 12. doi: 10.3389/fpls.2021.711896.

Ali, J. *et al.* (2024) 'Wound to survive: mechanical damage suppresses aphid performance on brassica', *Journal of Plant Diseases and Protection*, 131, pp. 1–12.

Ali, J., Sobhy, I. S. and Bruce, T. J. A. (2022) 'Wild potato ancestors as potential sources of resistance to the aphid Myzus persicae', *Pest Management Science*, 78, 3931–3938.

Bergström, G. (2007) 'Chemical ecology= chemistry+ ecology!', *Pure and Applied Chemistry*, 79(12), pp. 2305–2323.

Dicke, M. and van Loon, J. J. A. (2014) 'Chemical ecology of phytohormones: how plants integrate responses to complex and dynamic environments', *Journal of Chemical Ecology*, 40(7), pp. 653–656. doi: 10.1007/s10886-014-0479-0.

Dyer, L. A. *et al.* (2018) 'Modern approaches to study plant–insect interactions in chemical ecology', *Nature Reviews Chemistry*, 2(6), pp. 50–64.

Hartmann, T. (2008) 'The lost origin of chemical ecology in the late 19th century', *Proceedings of the National Academy of Sciences*, 105(12), pp. 4541–4546.

Mbaluto, C. M. *et al.* (2020) 'Insect chemical ecology: chemically mediated interactions and novel applications in agriculture', *Arthropod-plant Interactions*, 14, pp. 671–684.

Meiners, T. (2015) 'Chemical ecology and evolution of plant–insect interactions: a multitrophic perspective', *Current Opinion in Insect Science*, 8, pp. 22–28.

Millar, J. G. and Haynes, K. F. (1998) *Methods in chemical ecology volume 1: chemical methods*. Springer.

Mithöfer, A., Boland, W. and Maffei, M. E. (2009) 'Chemical ecology of plant-insect interactions', *Molecular aspects of plant disease resistance*. Wiley-Blackwell, pp. 261–291.

Prokopy, R. J., Cooley, S. S. and Phelan, P. L. (1995) 'Bioassay approaches to assessing behavioral responses of plum curculio adults (Coleoptera: Curculionidae) to host fruit odor', *Journal of Chemical Ecology*, 21, pp. 1073–1084.

Roberts, J. M. *et al.* (2023) 'Scents and sensibility: best practice in insect olfactometer bioassays', *Entomologia Experimentalis et Applicata*, 171(11), pp. 808–820.

Siljander, E. *et al.* (2008) 'Identification of the airborne aggregation pheromone of the common bed bug, Cimex lectularius', *Journal of Chemical Ecology*, 34, pp. 708–718.

Tholl, D. *et al.* (2006) 'Practical approaches to plant volatile analysis', *The Plant Journal*, 45(4), pp. 540–560.

Tholl, D. *et al.* (2021) 'Trends and applications in plant volatile sampling and analysis', *The Plant Journal*, 106(2), pp. 314–325.

Van Tol, R., Visser, J. H. and Sabelis, M. W. (2002) 'Olfactory responses of the vine weevil, Otiorhynchus sulcatus, to tree odours', *Physiological Entomology*, 27(3), pp. 213–222.

Weeks, E. N. I. *et al.* (2011) 'A bioassay for studying behavioural responses of the common bed bug, Cimex lectularius (Hemiptera: Cimicidae) to bed bug-derived volatiles', *Bulletin of Entomological Research*, 101(1), pp. 1–8.

Index

Brassicaceous plants, 105, 16
Brassinosteroids (BRs), 136–137
Bruchins, 102
Bugdorm bioassay, 162
Bursera, 59

C

Cabbage looper, 92
Chemical adaptation, 3
Chemical communication, 2, 7–8, 12, 32, 47, 94
Chemical compound, 1, 3, 4, 9, 16, 19, 65, 74, 143
Chemical cues, 3, 7, 8, 11, 12, 21, 34, 44–46, 85, 91, 94, 101, 127, 143, 160
  behavioural responses to, 85
  language of, 111
  in predator–prey relationships, 8
Chemical defences, 6, 55; *see also* Plant chemical defence
Chemical diversity, 87, 88
Chemical ecology, 1–2, 84, 93, 101; *see also individual entries*
  in agriculture, 125–126
    diversified agroecosystems, 138–143
    harnessing priming, 141–142
    insect herbivores, 126–128
    lures, 143
    natural enemies, 142–143
    phytochemicals, plant defence, 132–138
    phytohormones, plant growth, 128–132
    pollination, 142
    sustainable pest control, 138–143
    traps, 143
  applications
    conserving endangered species, 7
    disease vectors, 9
    in new pharmaceuticals, 7
    pest control strategies, 6
    in pharmaceuticals industry, 7
    understanding interactions between organisms, 6
  behavioural bioassay

  olfactometer bioassay, 163–167
  wind tunnel bioassay, 162–163
chemical communication, 7–8
chemical interactions, 161
disciplines
  biology, 4–5
  chemistry, 4
  ecology, 4, 5
at ecosystem level, 6
electrophysiology, 171
entrainment collection, 167–169
evolution, 2–3
historical context, 2–3
importance, 5–6
and insect–plant interaction, 9–12
insect–plant interactions, 161
at molecular level, 6
at organismal level, 6
performance bioassay
  BugDorm bioassay, 162
  clip-cage bioassay, 161–162
principle, 6
studies in, 3
volatile analysis, 169–171
Chemical interactions, 2, 3, 6, 8, 144, 161
Chemical messengers, 16, 21, 111
  pheromones as, 45, 46
  semiochemicals as, 23
Chemical signals, 1, 2, 5–8, 16, 23, 110, 111, 160
Chemical-to-electrical transduction, 37
Chemist, 4
Chemistry, 4, 7, 58, 101, 114; *see also* Plant chemistry
Chemoreception, insects
  anatomical adaptations, chemical sensing
    chemosensory information, 44–45
    morphological variations, antennae, 44
    mouthparts, 44, 45
    sensory structures and distribution, 43–44
  central processing of chemosensory input, 37
  chemical detection structures, 34–35
  chemical senses, 35–36
  chemical-to-electrical transduction, 37

Printed in the United States
by Baker & Taylor Publisher Services